CREATIVE PHOTOCOPYING

Creative

Photocopying

Using the photocopier for crafts, design and interior decoration

Stewart and Sally Walton

Photographs by Michelle Garrett
Illustrations by Jane Hughes

AURUM PRESS

First published in 1997 by Aurum Press Ltd
25 Bedford Avenue, London WC1B 3AT

Design: Hammond Hammond
Project Editor: Judy Spours
Picture Research: Emily Hedges

A Catalogue record for this book is available from the British Library.

ISBN 1 85410 476 4

Colour origination by Dot Gradations
Manufactured in China by Imago.

Contents

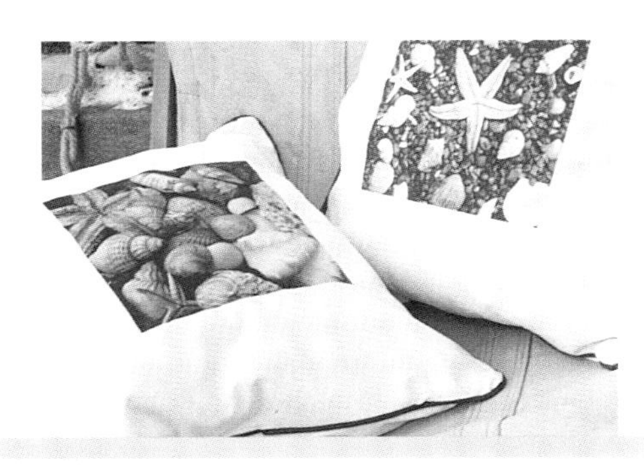

Introduction

EVERYONE KNOWS WHAT a photocopier does, but not everyone realizes what a photocopier can do. This book aims to inspire people to think differently about those great, grey boxes and to see them as remarkable tools to be used creatively, for all manner of arts and crafts projects. Whether you want to think big and turn your living-room into a replica of a Greek temple with photocopied columns and pediments, or whether you just fancy making a few unusual fridge magnets, you will find something here that persuades you to try your hand. But before turning to our inspirational projects, it is worth outlining the basics of photocopier design.

HOW A PHOTOCOPIER WORKS

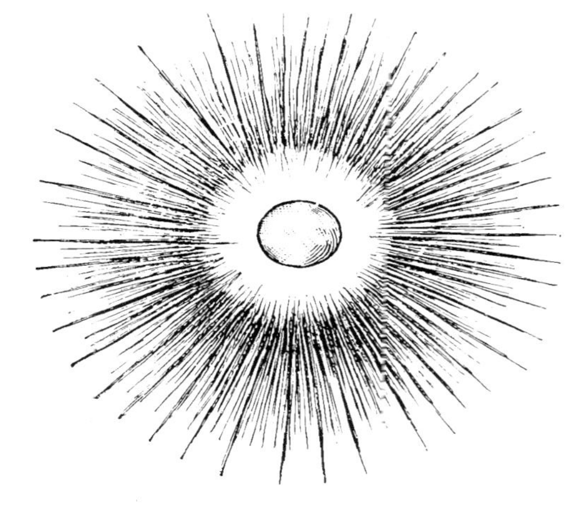

The word 'photo' comes from the Greek *phos*, meaning light, and as it is light that creates the copies, photocopying is a perfectly descriptive term. A piece of printed paper or other medium is placed on the glass inside the copier and the lid is lowered. What happens next, from our viewpoint, is that we hear some clunking and whirring noises – and may even see some brilliant light around the edges of the lid – then exact copies of our original come whooshing out of the side of the machine.

What actually happens is that inside the machine there is a surface coated with toner, a very fine powder. When the start button is pressed, a bright light shines through the image placed on the glass (some machines have a sliding scanner and others work by a single flash) and where it hits the surface, it knocks out particles of toner, leaving a positive impression of the image on its surface. Both the surface and the toner are electrically charged, creating an effective magnetism that causes them to cling to each other. A clean sheet of paper then passes over the surface and the electrical charge is dissolved, causing the toner to transfer onto the paper to make the copy. The surface is then clean and ready for a new image. Magic!

A colour copier works in the same way, but each copy passes over the printing surface four times. Full-colour printing is based on a system called C.M.Y.K. – which stands for Cyan (blue), Magenta (red), Yellow and Key (which essentially means black). All standard printed colours are made up of these four colours in different percentages. When a colour copy is made, each colour is printed separately, superimposing one over another to produce the full-colour image, which is why it takes a bit longer than a black-and-white print.

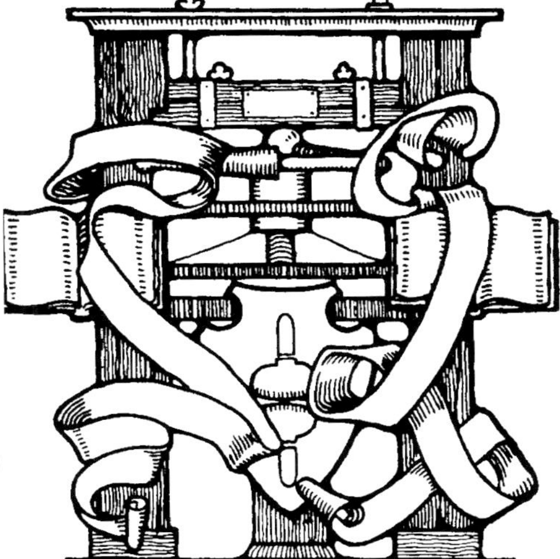

The quality of colour photocopying is improving all the time and will continue to do so as the technology wagon goes thundering along into the future. Colour copiers employ digital technology. Digital scanning means that an image is broken up into a grid of minute squares, each one of which is printed as a perfect individual. Each square is called a pixel, and the more you have (and therefore the smaller they are) the finer the quality of your print. The older, dot matrix system creates a less perfect image because a grid of dots will always have spaces between their circular forms. A digital copier can also isolate and print designated areas of an image or calculate the scale of enlargements needed to fill an area, meaning that all the things that we used to do manually, in advance, can now be done on the copier at the touch of a button. And the quality of these functions is improving all the time. Copiers can now even be linked to a computer to generate images without the need for a paper image under glass and flashes of light. But in this book we are are playing with paper, scissors, glue and colour, and having a bit of fun with the available technology to be found in copy shops everywhere.

USING YOUR CHARM

Most of the work that operators handle in photocopy shops is straightforward and there is little doubt that in stretching the possibilities creatively we need their help to get the results we are after. If you make complicated demands but still pay the same price as someone who simply wants a hundred copies of a newsletter, you can see why being creative doesn't make economic sense for the copy shop. You will need to make a friend of your photocopier operator, capture his imagination and involve him in your project.

If you are met by a blank stare of incomprehension when you explain what you're up to, it may help to take this book with you. If all else fails, it is time to move on to pastures new. If you can gain access to an art college, you will find that they have both black-and-white and colour copiers and will not be so surprised by an outlandish request.

We suggest going along to the smartest, most professional looking photocopy shop in town, which is where you will find the latest and best equipment, in top condition, and operators who will have been trained in both customer service and photocopy know-how. The only problem that you may then encounter is that, with a shop full of customers, staff may not have time to deal with complicated requests. Be understanding and suggest leaving the job with them to be collected later. In the process, launch an irresistable charm offensive in order to achieve your photocopying needs.

THE COPYRIGHT FACTOR

When photocopiers first appeared on the scene, copyright infringement was an immediate problem, but not on the scale that it is today. The quality of reproduction was poor and for most documents it was not convincing enough for a copy to pass as an original. The advent of plain-paper copiers had a big impact because they marked an end to the distinctive look of earlier copy paper. When photocopies began appearing on paper of all kinds and colours, they lost their distinctive shiny surfaces and became a lot harder to identify as different from originals.

Things have changed so much, especially with the introduction of colour copying and the digital revolution. Machines have become so technically brilliant that they can make copies that are almost indistinguishable from originals. This opened up a world of opportunities to the less than honest members of society – a potential forgers' paradise. Documents, photographs, identity cards – even bank notes – could be copied and passed off as originals. Something had to be done.

When you visit a copy shop now you will see a notice relating to the copyright laws, and there may also be a plaque fixed to the machine that draws your attention to the fact that it is forbidden to photocopy official documents and government papers. In the U.K., copyright does not need to be applied for. Additionally, for every piece of art, illustration, photography or written work, the copyright automatically rests with its originator.

The overriding advice of bodies governing copyright seems to be that you should ask for permission before you photocopy anything at all. The publisher of a magazine, newspaper or book should be approached first to find out who owns the copyright to the image you want to copy. It is highly unlikely that the answer will be straightforward. A publisher holds the copyright for a publication as a whole, but individual copyrights are owned by the originators of each piece of writing or of each image. Let's say that you are interested in copying a photograph. If the photographer is credited in the publication, you can try to trace him or her. Perhaps he has an agent that you can apply to. You can then ask permission to take a photocopy of the photograph and explain that you are not going to use it commercially but intend to paste it on your wall or onto a shoebox in the privacy of your own home. Well, one or two people with a lot of time to spare may find your efforts to stay within the letter of the law both charming and laudable. The majority, however, will wonder why on earth you are taking up their precious time and tell you exactly what to do with your shoebox.

So what is an honest citizen to do? All the material that appears in the source pages of this book can be copied as many times as you like because it is old enough not to be subject to copyright laws. There are also source books published by Dover Books Inc. from which any ten images can be taken without copyright permission being sought. You can also copy up to five per cent of any book without incurring copyright, as long as you have sought permission from the publisher.

After this you are on your own. Technically speaking, each time you make a copy of somebody else's work you are breaching their copyright, but, realistically, who is to know? This is the grey area. If you photocopy something from a book, make it into a frieze to stick on your living-room walls and it looks so absolutely marvellous that a magazine features it, the original artist, designer or photographer may recognize it. Most people would react with pleasure and pride that their work had been so good as to inspire further creative activity. Some may not take so kindly to it, so discretion is advocated. If you are in any doubt, then choose images that are not subject to copyright, those where the author or artist has been dead for seventy years, at which point the copyright expires. There is a 'special usage' clause that allows for photocopies to be made use of in schools, for private study purposes and for personal record-keeping. This is the clause which we think should be evoked to cover the type of copying we are suggesting in this book. One problem is that legislation has not kept pace with technology; and now that we can have colour copiers and scanners in our own homes, there is even less likelihood of enforcing the copyright laws.

You may find that the operators who makes your copies will refuse to copy certain things. They are acting within the law, which you should respect. There may be a company code of practice that forbids the copying of particular images – postcards of famous paintings, for instance. They may, however, agree to copy pictures of paintings from a book for study purposes. Their attitude is usually flexible as long as they are convinced that the copies are not part of a commercial venture. The latest digital machines have an in-built recognition system that will automatically throw out the colour balance on anything resembling a bank note, plus an identifiable, unique matrix of dots that make copies traceable back to the one copier that printed them. We think that we should be able to use photocopiers creatively as long as that does not involve the commercial exploitation of someone else's work.

DESIGNING

The most basic way to use a copier for designing is to make copies of images and type, then cut them out and rearrange them to put a message across. A computer can be used to generate different typographical fonts (styles of lettering) to be pasted up with your image, or you can photocopy lettering from a type source book and enlarge it to the size you need on the copier. This is useful for making up flyers, posters and notices for clubs or schools or for making your own greetings cards. Art and design supply shops sell type books, but before buying one you may like to ask at your local library, where they should also have photocopying facilities.

MAKING ENLARGEMENTS

Photocopiers can do a lot more than reproduce exact copies from originals. They can also enlarge or reduce a whole image or a selected area of it. As an image is enlarged, its quality changes –– a solid line at one size can change into a series of dots, dashes and squiggles when it is blown up. This is not always undesirable; in fact, it can look very arty when combined with some nice, crisp type. Reducing the size of an image can improve the line quality: a very loosely drawn image can tighten up considerably when brought down in size.

Explaining the size of enlargement required to the photocopier operator is easiest if you simply tell him the measurements that you want the image to have. For example, if you start with a 4 x 4 inch square and you want it to be twice as big, then ask for 8 x 8 inches. The percentage of enlargement for this, technically speaking, is just 40%, the result of percentage calculations which cannot be taught in detail here. Even more frustration is caused by asking for an image to be 'double the size' – in photocopy speak, this means double the height *and* length, making the area actually four times bigger. More accurately, calculate the percentage increase you require using the method described with our illustrations on page 12 and tell this to the operator, who can translate it for the photocopier if need be.

If you want an enlargement to cover a big area – for instance, double poster size – the copier will calculate how many A3 sheets are needed to fill the area and divide up the image accordingly. In doing so, it will automatically include an overlap, known as 'tiling', so that the re-assembled image can be pasted together with no visible seams. If you have ever watched the experts pasting up billboards, you will understand how 'tiling' works. If not, we suggest you align the two prints, holding them together temporarily with two small strips of low-tack masking tape. Use a steel ruler and sharp scalpel to cut down the middle of the overlapping edges. Remove the waste strips and the tape. The two prints can now be pasted down to make a continuous image.

FORMAT

Photocopiers are designed to take paper sizes in the metric 'A' range (the closest U.S. equivalents are eight and a half inches by eleven inches for A4 and eleven inches by seventeen for A3) and it may be important that your image can be adapted to fit this format. Most copiers do A4 (297 x 210 mm) and A3 (420 x 297 mm). A5 is half of A4, so the economical way to make this size is to line up two copies alongside each other and copy onto A4, then separate them.

REVERSING AN IMAGE

If you are making a colour copy which includes lettering – to transfer onto fabric, for example (see Chapter Three) – the image will first need to be reversed so that the type does not appear as a mirror image. This can be done at the touch of a button on a colour photocopier and there is a way of doing it on a black-and-white machine as well. The image is first copied onto a sheet of clear acetate, which is physically turned over, then copied again, this time onto paper. Copy shops usually have a supply of acetate; otherwise, it can be bought from art and design supply shops.

PAPER

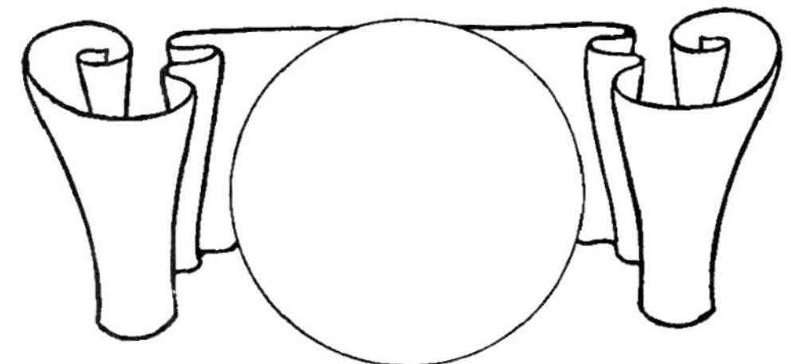

Photocopiers take many diferent types of paper, but some are unsuitable. The copy shop will have a selection of different types and colours but they are usually happy to use your own paper, unless they feel it could damage the machine. Problems come when textured or recycled, rough papers with loose particles of leaves, fibres or chippings leave bits behind inside the photocopier and could even cause it to catch fire. Paper with a smooth surface works best, but you can also copy onto thicker-textured watercolour paper, which is made from a very fine pulp.

Fancy papers can be very expensive, so if you have any doubt

ADAPTING AN IMAGE TO FIT A STANDARD FORMAT

Take an A4 or A3 sheet of paper. Put your image – here we use a colour-print photograph – in the lower, left-hand corner. Place a ruler diagonally over the paper and photograph to make an imaginary line from the lower left corner to the upper right corner of the paper. If the ruler also passes through the top right-hand corner of the photograph, your image is the same shape and proportion as the sheet of paper. In this case, you can simply enlarge your image on the photocopier until it fills the whole sheet.

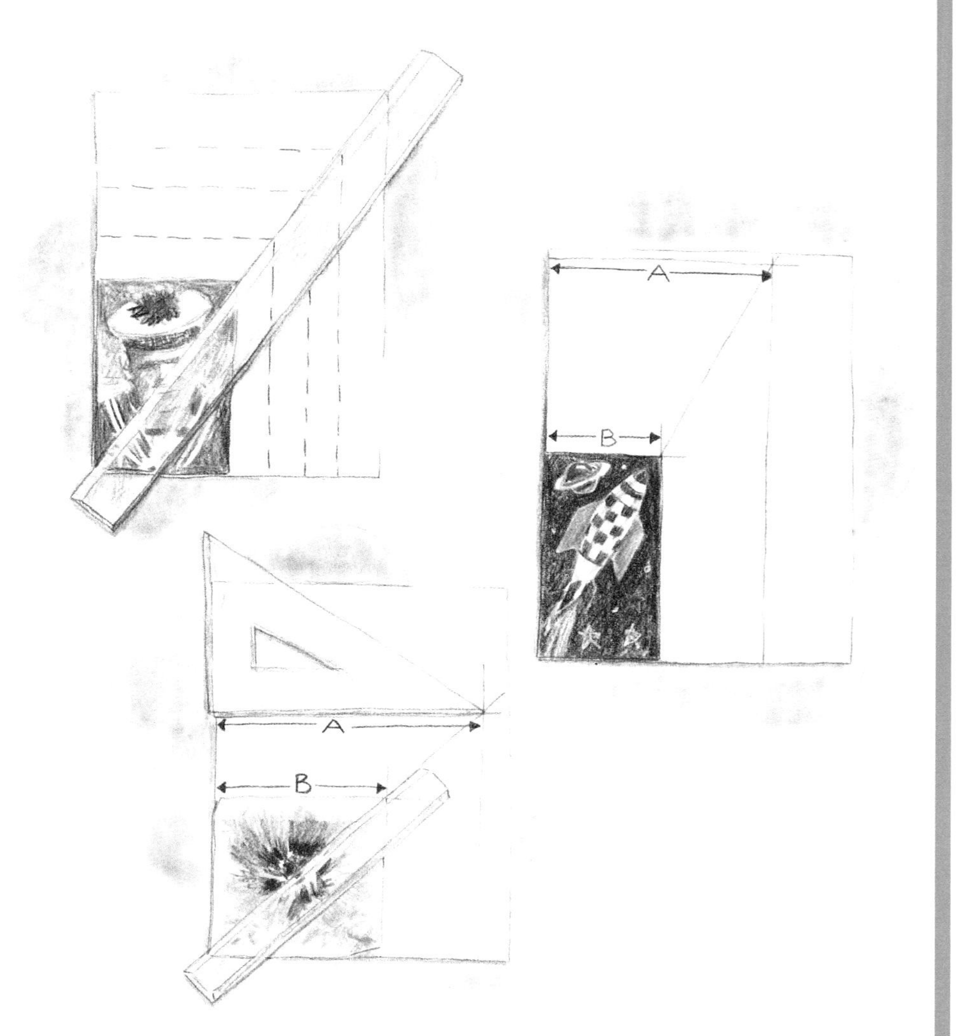

However, if the ruler passes through the top edge of the photograph at a different point, mark this on the paper. You can then either trim your photograph to size so that its top right-hand corner coincides with the pencil mark on the paper (the image may even look better with some of the background area cut away) and photocopy as before; or retain the full photograph and mark the largest area possible for an enlargement by placing the ruler across the photograph's own diagonal.

To work out the percentage enlargement you then need to ask for at the copy shop, measure the drawn-up sides 'A' and 'B' (see illustrations), divide the length of 'A' by that of 'B' and multiply the result by 100. This will give you the required percentage enlargement of side 'B' to ask for. (E. g., 25 ÷ 20 x 100 = 125 per cent of side 'B'.) To reduce the image, divide 'B' by 'A' and multiply by 100 (20 ÷ 25 x 100 = 80 per cent reduction of side 'A'). A calculator makes this simpler than it sounds.

about whether or not a paper will be suitable, take a sample to the copy shop and get their opinion first. A shop will be understandably reluctant to risk machines for the sake of one sheet of Javanese mango-fibre paper and will just refuse if it looks dubious.

One last point to bear in mind about paper is that it needs to be trimmed to an `A' format (or the U.S. equivilent) to fit into the photocopier.

CUTTING OUT

The choice is between cutting with scissors or with a blade, and some people swear by one, while others find that they cut more accurately with the other. A scalpel or craft-knife blade is ideal for cutting straight lines, with the aid of a steel ruler. The best surface for this is a cutting mat, which has a self-healing surface so that the blade leaves no grooves. Cutting mats are conveniently marked with a grid of vertical and horizontal lines, which makes short work of cutting right angles. If you find a scalpel uncomfortable to hold, try binding the handle with an elastoplast (Band Aid) – this will give you a non-slip, soft pad to hold instead of a piece of narrow steel or plastic. It is very important to renew scalpel blades regularly as a blunt blade will snag and spoil the clean edge of the paper.

Most of us are more familiar with scissors and therefore feel in more control of our cutting if we work in this way. If you choose scissors, have three different sizes. A large, long-bladed pair for cutting large areas; a medium pair for cutting smaller shapes; and a small, pointed pair for cutting out details. Always cut with the scissors at the angle that is most comfortable and move the paper to meet the scissors, not the other way around.

MEASURING

A long steel ruler marked in inches and millimetres can be used as a cutting edge and a measure. It is also convenient to have a short ruler when you are working on a small scale. A retractable measuring tape is best for measuring large areas like walls, furniture or fabric, and a clear Perspex (hard plastic) set square (triangle) with a 90/45 degree angle is helpful for checking verticals and horizontals. Use a spirit level when working on walls; you can either fix one to a length of batten with masking tape or buy a decorator's metal ruler with an integral level. There are two other very low-tech measuring methods that we find useful. The first is the plumb-line – a length of string with a piece of reusable adhesive (such as Blu-Tack) on one end and a key on the other. This is attached to the top of a wall and allowed to hang free to skirting-board height, to give a perfect vertical line. The other idea is even more basic. If you are using a repeating measurement, cut a strip of card or stiff paper the correct width and use it to mark off the distances. We used this when painting the red and white stripes to go with the children's room frieze on page 43. It may seem like stating the obvious, but it does provide a regular, accurate measurement and is easier than moving a ruler or tape measure along a wall.

ADHESIVES

We have advised using various different types of adhesive in the projects. The first is a simple glue stick. This is the ideal, no-mess method of sticking elements of a design onto a sheet of paper to make a photocopy. This would be the choice when you are arranging type with an image because all you need is a central dab of glue to hold each piece in place.

For a permanent paper-to-paper bond, needed for a project like bookbinding (see page 29), we suggest using an aerosol adhesive such as Display Mount or Art & Craft Adhesive. It should be applied in a light, even coating and will bond permanently if the surfaces are joined when the glue is wet. An aerosol adhesive such as Spraymount can be used to give a temporary bond, by allowing the glue to become almost touch dry before joining the surfaces. This is worth considering if accurate positioning is essential to your final image, as it will give you a chance to correct any slight errors, such as a line of type that is not quite horizontal. A word of warning – aerosol adhesives do tend to drift, so it is essential to protect your work area. A cut-away cardboard carton will make a very good, temporary spray booth.

All of the interior and furniture projects use wallpaper paste. The powder can be mixed up in small or large amounts, according to the manufacturer's instructions, and the paste is applied, ideally to both surfaces, with a brush. It seems to give paper a slightly stretchy quality when wet, which is useful when fitting the paper around an awkward shape like the handles in the 'knitted' tray project on page 96. The paste gives a very strong bond, and as it dries, the paper loses its elasticity, tightening up to give a smooth, taut surface.

Many of the projects employ Blu-Tack, the very versatile, pale blue, plasticine-like material. A paper image can be temporarily fixed in place with Blu-Tack so that you can see how it looks before permanently pasting it down. Blu-Tack is not available in the U.S.A., but double-sided masking tape can be used in its place. In fact, anything that can be stuck on and then removed without leaving a mark will do; a thin coating of Spraymount is suitable on some surfaces

We hope to capture your imagination in the pages that follow. The projects vary from very simple to quite complex. Some can be done in an instant, like the laminated table-mats, where the only creative input will be your choice of image; others, like the fireplace, take much longer and require thought in the selection and design process, as well as skill and patience to complete the project. The results in either case should be a delight, and we are sure that once you try one project you will be inspired to go on to another and many more of your own invention. At the end of the book, we have included thirty-two pages of images to photocopy for decorative schemes. Many of the patterns and motifs we have used in our projects are on these sheets, together with numerous others, arranged under subject headings, which we hope will inspire you. The sheets can be detached for easy use, or you can take the whole book along to the copy shop.

We hope that you will have as much fun as we did with photocopying projects. We actually found that our output of new ideas increased as we experimented with the photocopier and we continue to discover new functions and picture sources that give rise to yet more innovative design possibilities.

Chapter One

Stationery

PHOTOCOPIERS WERE INVENTED to copy text – all the documents and letters that once were either copied using carbon paper or typed onto a special waxed sheet for a duplicating machine could now be copied electrostatically, at the touch of a button. This was the beginning of a printing revolution, which had the unexpected social spin-off of freeing typists from their desks. Photocopiers were large and fairly noisy machines that needed their own rooms, to be visited by people from all departments. Queues formed and friendships, even romances, developed as a result. Where would we be without the photocopier?

In this first chapter, we look at creative ways of photocopying to make your own letterheads, labels, gift wraps, book and box covers. Most of us lead busy lives in which we tend to compartmentalize work, leisure and creative pursuits, but most of these ideas invite you to cross over between the preoccupations. When you need to write letters, for example, you buy stationery; in the same way, when you give a gift, you buy wrapping paper. But why not make them instead? That way you enjoy the creative

process, as well as extending the pleasure of receiving a gift or a letter by making your offering a bit more personal.

Whilst many of us have access to computers that can print out different typefaces, scanners that can copy photographs or images are still the province of graphic designers. This is where the photocopier comes into its own. You can use the computer to select your typeface, print it out in the size, style and arrangement that you want, then marry this with the image of your choice using a photocopier. As a design tool, it will give you instant feedback at a very low cost, so you can change your design until you perfect it. And if you don't have a computer, there is rub-down lettering, or photocopied enlargements can be taken from a type source book.

Once you have grasped the idea of using photocopied images to customize paper for use at home, the possibilities seem endless. Book covers, cassette cases, box files, note pads and storage boxes can all be given this decorative treatment. Why not make your own in-house recipe book, mixing handwritten instructions with photographs or illustrations? Providing that your cookery is admired, it could even make a great present for friends, especially if you follow our project instructions to bind it up in an appropriate cover.

In each of these opening projects we try to persuade you to think of the photocopier as just another tool to be used with creativity and originality. Wherever you live, there is likely to be a photocopy facility nearby, so why not have a go? Do experiment: it doesn't cost much and you can have a lot of fun.

LETTERHEADS

One of the quickest and easiest projects that you can do on a photocopier is to make your own letterheads. You don't have to commit yourself to buying a box of paper – you can print just one sheet for a special purpose or make several variations to suit the mood of your correspondence.

There are no hard and fast rules to follow, just choices to make. Think of the message that you want your paper to put across. The colour and weave of the paper that you choose can say as much about you as the image at the top, so rather than settle for standard copier paper, look around for something that expresses your personality – are you a satin smoothy or more of a natural, earthy personality?

The feel of a sheet of chunky recycled paper will never conjure up a vision of pampered loveliness, but it will make an impression on your local organic vegetable supplier! If you want to be taken seriously, then use paper with added weight – solicitors and

architects do so. For letters that are full of fun and gossip, choose paper that is light, bright and colourful. There is so much to choose from, and you need not feel limited to using writing paper – gift wrap or art papers can be cut down to size and the copy shop will probably be able to do this for you on a guillotine.

When it comes to the choice and positioning of the image, you will see from our examples that you can use anything that appeals to you and place it in the centre or off-centre. As you will be using a black-and-white copier, it is best to choose clear images to start with and to bear in mind that the type of paper you print on will have an influence upon the tone of the black – it looks quite different on smooth white than it does on a textured red paper.

Experimentation is the key, once you've selected an image. Make six copies – two of each in three different sizes – then cut them out and spend a little time moving them around the top third of a sheet of A4 paper to see how the spacing works. When you are happy with the way it looks, stick the various elements or single image down using a dab of glue and it will be ready for copying. Have a single copy made and check that it has worked out the way you planned – this is especially important if you are using special paper where mistakes could be costly.

It is a good idea to keep one perfect copy safely in a folder so that you can use it as your master copy and have more paper printed as and when you need it.

JAM LABELS

HOME-MADE PRODUCE is a treat (well, most of the time) and deserves to be seen, not hidden away in a cupboard with no identifying marks. If you are making jams or pickles, the chances are that you will offer them to visiting friends or give jars away as gifts. If you are smart, you will remember always to ask for the empty jars to be returned for next season's efforts, but you will seldom get the same jars back. You may find that when the time comes to make a new batch, you are faced with an assortment of ill-matched jars; but don't despair, you can turn them into a set as distinctive as any of the recognizable top brands – all you need is your own set of labels and covers for the lids.

The great thing is that you are free from all the constraints that dog the lives of commercial producers – you don't have to list the ingredients, weight or calorie content – all you really need is to identify the contents with the month and year in which they were made. This leaves you free to express yourself and design labels that say who made this delicious offering. Perhaps you will go for a pastiche of an existing label, or borrow a look that you have admired elsewhere – French country style, for instance. Or you could put your own photograph on the jar in a collage, surrounded by the featured fruit or veg. Let your imagination run wild.

YOU WILL NEED

- **Black-and-white photocopies of fruit**
- **Spraymount**
- **Strong, transparent craft glue**
- **Rubber bands and narrow ribbons**
- **Scissors**
- **Small sheet of cardboard**

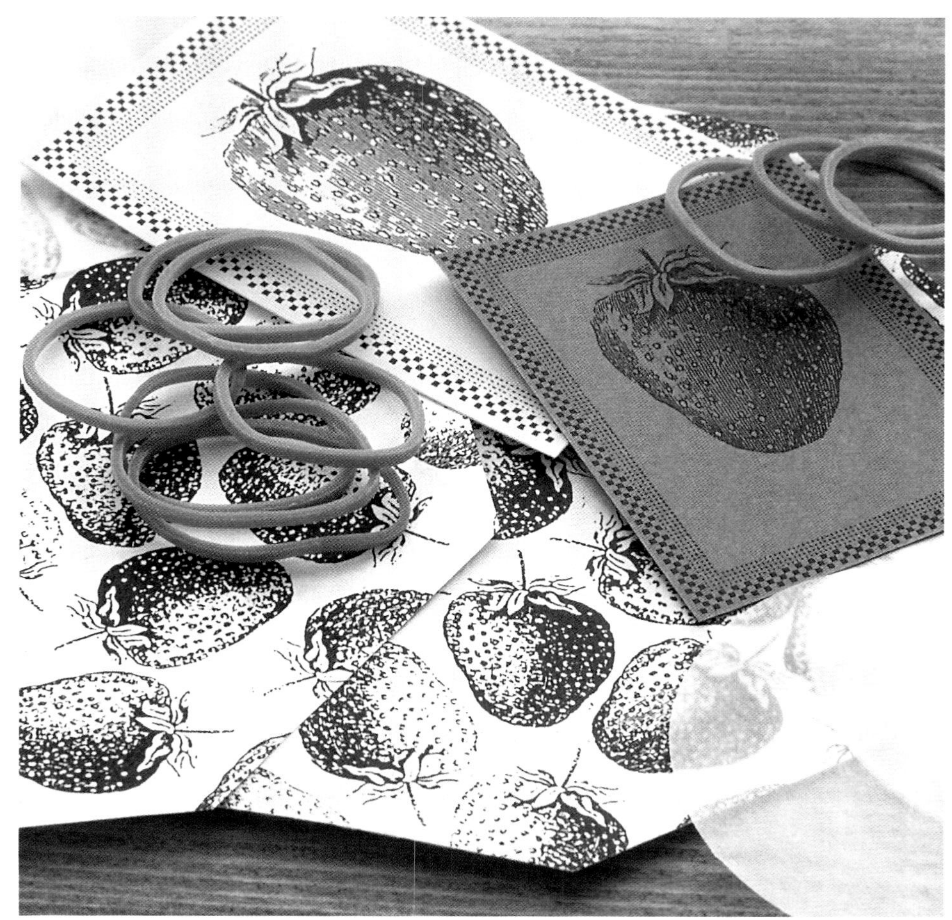

7/8/97

HOW TO DO IT

1

1 To make the label, photocopy the checkered border on Sheet 4 Measure the space that yoɹ have on the side of your own jar, because jar size does vary a lot. Have a reduction or enlargement made and trim the edges.

2 Enlarge the strawberry on Sheet 8 or another fruit so that it fits the frame, allowing a larger space at the bottom in which to write the date when the batch was made.

3 Cut out the stawberry and use Spraymount to stick it in the frame.

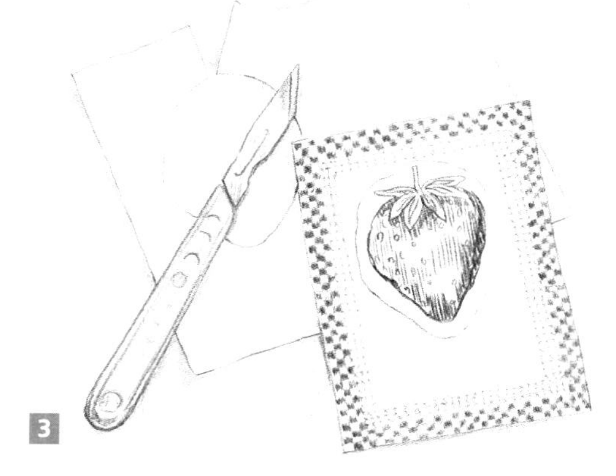

3

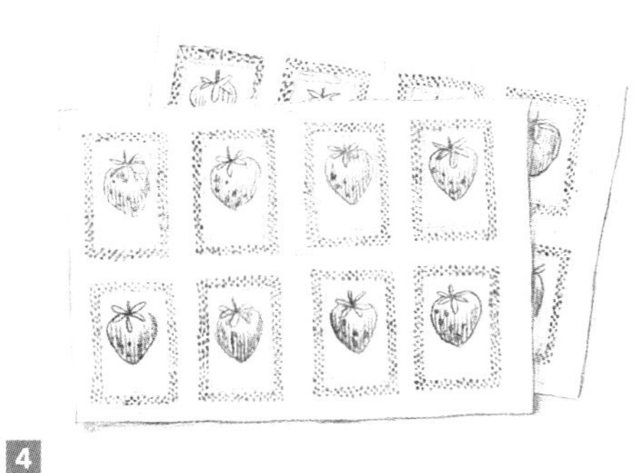

4

4 Photocopy the label twice, then copy the two next to each other; keep doubling up the copies in this way until you have the number of labels you need.

5 Use a strong transparent glue to stick the labels onto the jars.

6

6 To make the matching cover for the lid, have the original stawberry reduced in size, to half it's original height. Using the same method of doubling up on each copy, make 64 small strawberry prints to arrange on an A4 sheet. Use Spraymount to stick them to the paper, then photocopy the sheet.

7 Draw an octagon shape, measuring about 5 inches (12.5 cm) across, on a piece of cardboard. Cut it out to use as a template.

8

8 Draw and cut out as many tops as you need, then line a ruler up to opposite corners and made a crease about an inch and a half (4 cm) long. Do this by folding the paper up against the ruler edge and running your nail along the underside. Do this all around the edge and the covers will pleat up very nicely – lifting your jam pots into another realm altogether!

9 Hold the covers tightly in place with rubber bands, then hide the bands with narrow ribbons.

GIFT WRAPS

IN THE GIFT wrapping business fashions change fast, and the paper that you notice one month and return to buy the next may have vanished forever. New ranges are constantly being introduced – one moment big florals are everywhere and the next you can only find stars. Worse still are the thousands of variations on cute teddy bears or the latest blockbuster cartoon character. One solution would be to buy the paper you like when you see it and stash it away until you need it, but another is to photocopy images you like and make your own wrapping paper on a machine that produces A3-size prints.

Black-and-white prints are best because the cost of an A3 colour copy is too high to justify using the process for anything other than a very special occasion. Black-and-white certainly doesn't have to stay that way – a plain paper copier can print onto most papers, so you can start with a sheet of plain-coloured gift wrap and add your images, or use white and colour it yourself using marker pens, paints or crayons.

The papers that we have made up here are the absolute tip of the iceberg – but they will give an idea of different source material that you can use. The lace pattern was made by placing a sheet of (dare I say it?) plastic lace on the copier, backing it with a sheet of black paper and photocopying onto cream paper. The calligraphy is an enlargement of an old letter found in a history book and the sun motif appears in a source book of copyright-free images for graphic designers.

Just think: any printed image can be photocopied – plane tickets, concert programmes, CD covers, menus, newspaper cuttings, adverts, wrappers – all of which provide rich source material that can be used to make the most stunning one-off gift wrappings. Your biggest problem will be finding a gift that won't be outshone by the paper it's presented in!

YOU WILL NEED

- **Selection of paper – brown parcelwrap, plain coloured paper, etc.**
- **Selection of images to reproduce**
- **Ruler**
- **Set square (triangle)**
- **Spraymount**
- **Sheet of A3; black backing paper**

SUN PAPER

This motif appears in the book on Sheet 32, so the first step is to take a copy. Ask the print shop to use the copy to make a multiple print on A4 paper. This can only be done in one direction, so from this you will get a strip of spaced images. Have five copies made. If you are working on an old copier it may not have this facility, in which case make four copies, snip them out and arrange them on the copier and make another three copies – what you are doing is getting value for money and cutting down the number of prints you need to fill the paper. Cut out all the sun motifs and arrange them on an A3 sheet of paper (if you would like the motifs to be larger, you could do this using fewer of the motifs on an A4 sheet then have it enlarged to A3 size). Use a ruler and a set square to check that the lines are straight. Spraymount them in position, then have this artwork copied onto your chosen paper.

CALLIGRAPHIC PAPER

Select your text and chose the best section of it to enlarge. Try a few different sizes until you find the one that looks best. The object is to turn the type into something decorative rather than legible. There is no more to it than this. You can vary the effect by your choice of paper – something slightly transparent that imitates vellum can look like an authentically old manuscript, whereas bright colours can look bold and contemporary.

PAPER LACE

As we have already admitted, this is actually a piece of plastic lace (bought in an Italian market, however, so it does have some aesthetic credibility), but you could use any piece of hand-or machine-made lace. You can always find

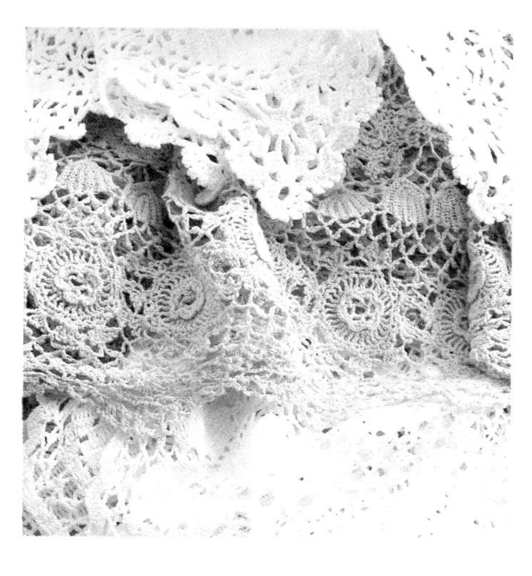

doilies and tray cloths in junk shops going for a song, because so few of us have a use for them these days. Just one small, lace-edged coaster can be copied and repeated in the same way as the sun motif, or you can take a print from a whole curtain panel. Adapt this idea to use another distinctive open-work textile like crochet, macramé or tatting. The lace is put directly onto the glass, then covered with a sheet of dark paper which prints as black.

BOOKBINDINGS

SIMPLE BOOKBINDING IS an extremely satisfying task. You start with a few off-cuts of mounting board, a photograph and a pad of paper, do a bit of measuring, sticking and smoothing and suddenly you have a neat and entirely exclusive little notebook. When you have followed the instructions once, you will be able to apply this technique to make a whole range of matching desktop accessories. Don't feel limited to the desk area, either; you could easily make your own photograph albums, visitors' books, school folders or binders for sets of magazines.

You do not need a lot of specialist materials – just a strong aerosol adhesive, like Display Mount or Art & Craft Adhesive (not the type that allows for repositioning, as it will not be strong enough for bookbinding). Mounting board is an ideal weight for the covers and you may be able to buy a bag of off-cuts from a local picture framer – when they cut window-mounts, the middle section is often treated as waste, but is just right for bookbinding. Printers and copy shops always have pads of paper for sale that they make up from the off-cuts of large print runs – these come in all shapes and sizes and can be cut to the size you need on their power guillotines.

You are not limited to an enlargement of a photograph: once again, anything that appeals to you and will fit in a photocopier will do. The ivy we used grows on our house and the pebble wall is just down the road – we pass them everyday, along with thousands of other textures and patterns that would make equally stunning covers for books. Sometimes just one small section of a photograph has a perfect arrangement of colours, and this can be selected and enlarged to make a wonderful abstract print. Look around and become inspired.

YOU WILL NEED

A photograph or any printed image of your choice
Mounting board – the type used by picture framers, sold in art shops
Aerosol adhesive such as Display Mount or Art & Craft Adhesive
Ruler
Set square (triangle)
Contrasting paper for the spine (optional)
Lining sheet of paper to fit inside the cover
Pad of paper – make the bindings first, then have a pad cut to fit.
Spoon to smooth the glued flaps flat.

1 Have an A4 colour enlargement of your photograph made and trim it to the edge of the print. The photographs used here were taken very close-up to make the most of the textures of the pebble wall and ivy.

2 Place the print face down on your work surface and draw a pencil line down the middle and another three-quarters of an inch (2 cm) up from the bottom as your base line.

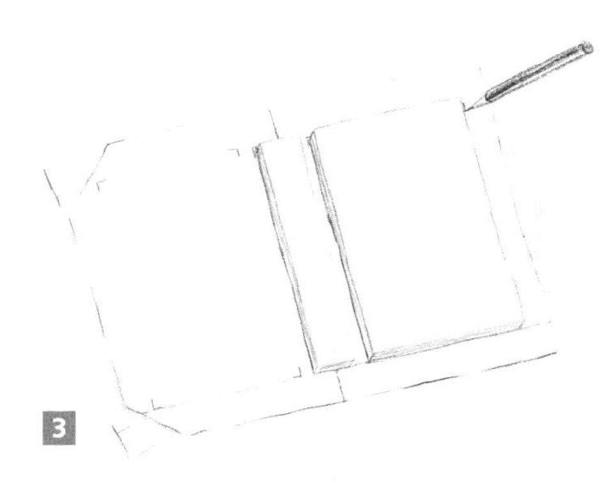

3

3 Cut matching pieces of card of the same measurement for the front and back, and one spine section to accommodate the width of your eventual paper pad. Lay the card pieces on top of your print, resting them on the base line and with the spine centrally positioned along the centre line. Allow a three-quarters of an inch (2 cm) overlap all around and a quarter inch (5mm) gap on each side of the spine. Use a pencil to mark the corner positions of the three pieces, then remove them. Now trim away the corners of the print to within an eighth of an inch (3 mm) (or the depth of your mounting card) of the four outer corner pencil marks. It is important to leave this space to give a neat finish when the flaps are folded over.

4 Place the print on a sheet of newspaper and apply an even, light coating of spray adhesive. Stick the three pieces of card down in the positions you have marked.

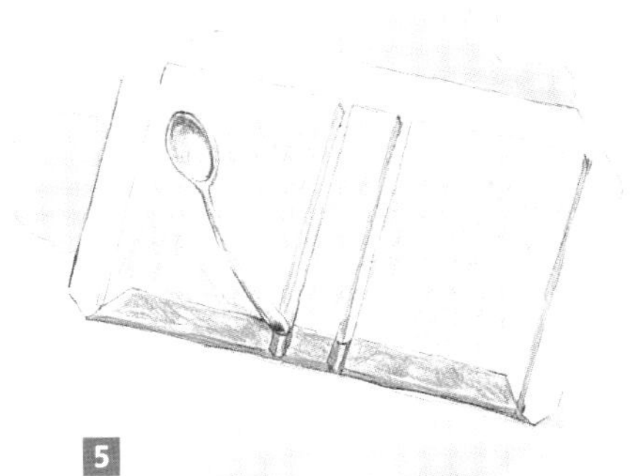

5

5 Use something smooth and rigid, like the handle of a spoon, to fold the side flaps up against the edge of the mounting card, then flatten them over and smooth them down.

6

6 Now is the moment when you understand the significance of that eighth of an inch (3 mm). Push this slight overlap down into the edge of the board, to follow its shape. This is the secret of neat corners. Fold and stick the other flaps down.

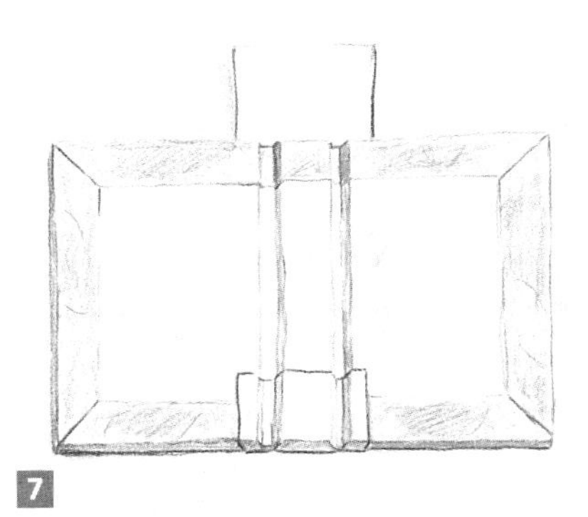

7

7 Cut and glue down the contrasting spine strip, allowing a good two and half inches (6 cm) overlap onto the inside.

8

8 Cut a lining sheet a few millimetres smaller than the cover, spray with adhesive and stick it down to cover all the overlaps. Fold the finished cover into a book shape.

9

9 Take the binding to the copy shop and ask them to trim a note pad to fit inside, then stick the last page to the inside back cover, using spray adhesive.

VIDEO STORAGE BOXES

VIDEO TAPES SEEM to have the ability to breed, and unless you have an effective storage system they clatter about all over the place. Another problem is the way they all look alike without their covers; but don't despair, here is a simple, cheap storage idea that will sort out your tapes and look stylish at the same time.

The basic plan is to customize large shoe boxes by covering them with suitable film stills. The best boxes size-wise are the ones that contain men's training shoes or chunky boots – look for a friendly face in a sporting equipment store and ask for any spare boxes. A lot of people leave the box behind when they buy shoes and shops are pleased to make space by getting rid of them (have the measurements of a video tape handy so that you can check the fit).

Having acquired enough boxes to house your video library, your next stop is the public library in search of covering material. There may be a book on old film posters or arty foreign films. If you have a penchant for Westerns and fifties horror films, our box is bound to appeal. The images you choose can reflect the sort of films in the box – cartoons for the kids' films, Fred and Ginger for the forties musicals, and so on. You could take a less literal approach and use associated imagery rather than film stills – especially for home recordings of cookery, wildlife, sports programmes and the like.

The next creative endeavour is to make a pattern of the box and collage all the photocopied material to fit the shape. A final photocopy is then made of the collage and this is used to cover the box – hey presto!

YOU WILL NEED

Several large shoe boxes
Photocopied images for your collage
Brown paper to make the pattern shape
Spraymount
Wallpaper paste and brush to apply it
Scissors
Scalpel
Acrylic varnish

1 Measure the sides of the box, adding on an extra 2 inches (5 cm) to allow for overlaps both top and bottom. Cut the shape from plain brown wrapping paper to use as a pattern. Measure the top of the lid, adding on the side measurements and an overlap of half an inch (13mm) all around the edge.

2 Select your images and have plenty of black-and-white copies made. Cut them out and arrange them on the paper pattern shape to make your photo-montage. Cut out different characters and place them in unlikely scenes — we have two guys approaching the corner from different directions, destined never to meet for the final shoot out!

3 Spraymount all the elements in position then have the collage photocopied again. This may take two A3 sheets or several A4 sheets, depending upon the dimensions of your box.

4 When covering the sides of the box, begin about 2 inches (5 cm) in from a corner, pasting both surfaces right up to the edge. Allow an equal overlap top and bottom. Smooth the paper up to the corner, then use the scissors to cut a narrow slot from the top, down to meet the corner – just the thickness of the cardboard – and fold the top overlap inside the box. On the bottom edge, the corner is mitred at 45 degrees, so snip out a triangular piece, then fold under and smooth flat. (This will not show, so slight imperfections are permitted.)

5 Continue pasting and wrapping around the box, but don't cut the corner slots or mitres until you reach them because the paper stretches a bit when pasted.

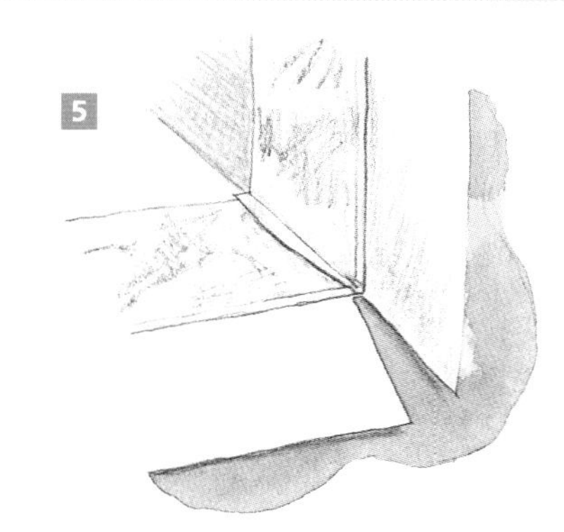

6 To cover the lid, make one A3 photocopy (ours is a colour copy of an old movie poster), place it face down on a flat surface and trim the paper so that there are even side and overlap allowances all around the edge. Place the lid in the middle and draw around it, then remove the lid and use a ruler to extend the lines to the edge of the paper, making a square at each corner.

7 Paste both surfaces, then smooth the paper onto the top of the box. Use scissors to trim away the corner waste, starting about a tenth of an inch (2mm) inside the square and slanting into the corner point. This will give you the slightest overlap and a very neat corner finish. Paste the sides and inside edge, then smooth the paper down and tuck it under the the lid and smooth flat.

8 For a durable finish, a coat of acrylic varnish can be applied when the paste has dried.

Chapter Two

Interiors

DECORATING INTERIORS with photocopies may initially sound a bit like shop window dressing or the sort of thing you might do for a party, but not for everyday life. But photocopying is in essence simply another way of putting patterns onto paper, no different in principle from wallpaper, which has been employed for centuries. We are not suggesting that you print rolls of wallpaper in your local copy shop. The staff might object to you monopolizing their machine for the length of time that would take. What you can do, though, is print borders, panels, friezes or individual motifs that can be applied to your walls using paste and topped with a coating of varnish to provide a seamless wall covering.

When you take the usual route and look through a wallpaper pattern book, the first things that are likely to strike you are how similar patterns are and how certain fashionable colour combinations dominate the range. At the top end of the market there is much more variety of design available, but the prices are astronomical. By using a photocopier, you can choose exactly which images you want for your walls, by-passing conventional wallpaper altogether. You may want something simple, like a bow or a flower to repeat along a border or an engraved sun motif to arrange in a grid as an all-over pattern. Both can be done on a black-and-white copier and you can add colour by hand. On the other hand, you may not want a conventional arrangement of images at all. You might instead add architectural features by enlarging pictures of columns, mouldings or carved panels for some instant *trompe l'oeil*. When the Grand Tour was all the rage in the eighteenth century, artists would travel to Rome and Pompeii and make painstaking drawings or engravings of the architectural wonders of Italy. These are all out of copyright now and provide excellent source material for introducing architectural splendour to your home. Have a look through the architectural history books in your local library for inspiration.

Most of our photocopied decorating ideas have an element of fun to them, but you can also make some seriously stylish statements in either black and white or colour. If you decide on colour, then we suggest using a black-and-white copier for all your initial designs and workings out: it is so cheap that you can afford to experiment and reject those ideas which don't work. Enlarge your designs to the correct scale and try them *in situ* to see if their proportions fit the room before making the expensive leap into colour.

Decorating with photocopies can change a room in an instant, and the transformation is likely to cost less than a single roll of good wallpaper. So, if you are feeling daring now but might become more traditional in six months' time, there is no problem – at this price you can afford to be fickle and change your style whenever the mood takes you.

SMALL PRINT ROOM

TRADITIONALLY, PRINT ROOMS are featured as part of a formal style of decorating, originating in grand, eighteenth-century country houses. Photocopied print rooms don't demand such high interior standards, but they can be devised to create something of a formal atmosphere even in a modest room.

If you have a favourite set of old engravings, you can photocopy them to go inside the frames we suggest. If not, a trip to an antiquarian book dealer or a local junk shop is in order to find suitable material to copy. So many obscure books from the nineteenth century have high-quality engravings – which was the nature of illustrations then – and they photocopy beautifully.

The frames should be varied in scale and design, so as to add further interest. You should have no problem finding suitable ones: some are provided at the back of the book on Sheets 4 and 5, and a number of source books devoted solely to frames are available from bookshops. The word 'frame' here is generic, embracing all manner of borders, swags and bows.

The traditional background colours for print rooms are yellow, terracotta or duck-egg blue, so if you are going for a traditional look, choose one of these. If, however, you're a decorating rebel, you'll have your own ideas about colour. In any case, the painted wall should be bone dry before you add the prints, so leave a day between these two stages.

Print rooms used to involve cutting a frame border with mitred corners and assembling the framed prints *in situ*. The photocopier once more comes into its own here, allowing us to cut out a lot of this work and stick prints of framed scenes or portraits on the wall as a single sheet.

YOU WILL NEED

- **A selection of engraved borders and frames – see Sheets 4 and 5 or try the sourcebook *Cartouches and Decorative Small Frames,* published by Dover Publications Inc.**
- **Engraved scenes or portraits to feature in the frames**
- **Masking tape**
- **'Antique' cream-coloured paper in A3 format**
- **Blu-Tack**
- **Wallpaper paste and a brush**
- **Clear matt varnish**

1

1 Make your selection of frames and borders. Enlarge some to fit an A3 sheet, but keep others small. Aim for variety in both size and shape. Cut out the plain, white centres of the photocopied frames, leaving spaces where the pictures are to go.

2 Find an old book with topographical engravings, portraits, etc. that will suit the mood of your room. Enlarge your chosen images on the photocopier to fit the various frames you have already copied. Some of the images may need cropping (cutting down) from the full image to fit the frames.

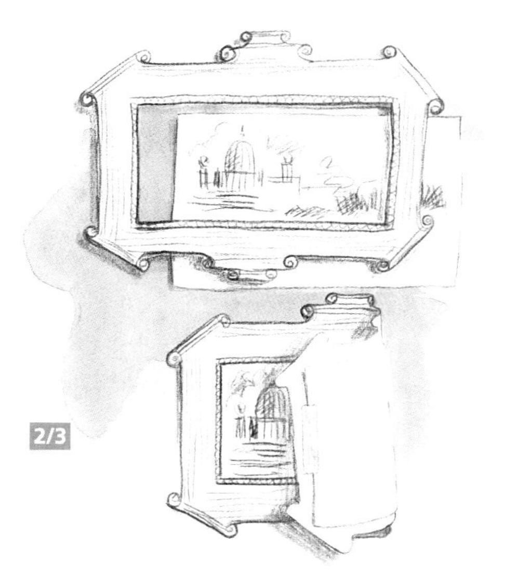

2/3

3 Use small strips of masking tape on the back to hold the photocopied prints in place in the centre of the photocopied frames, making sure that the prints slightly overlap the inside edges of the frames.

4 Photocopy the prints and frames together onto A3-format 'antique' cream paper. The smaller prints and frames can be photocopied two to a sheet on the A3 paper. Cut out the finished print-and-frame photocopies.

5 Plan the arrangement of the framed prints on the wall, using small pieces of Blu-Tack to hold the photocopies in place. Once you are happy with the design, paste up the prints using wallpaper paste.

4/5

5/6

6 When the paste is dry, apply a coat or two of clear matt varnish over the prints and frames to seal the photocopies and the edges.

AN UNDERWATER BATHROOM

PAINTING MURALS REQUIRES time and considerable specialist skill, but photocopied wall schemes are almost instant and the only talent needed is with a glue brush. The rest is pure playschool pleasure – colouring-in for grown-ups. You may not want an all-over pattern in the bathroom, but bath-time will be a whole lot brighter with shoals of reef fish swimming past; remember, too, that their direction can be changed by reversing the images on the copier. Be bold with colour: if you're lucky, this project may coincide with the showing of a wildlife documentary about the Great Barrier Reef or the Red Sea; if not, a visit to a tropical fish aquarium will remind you that nature is the wildest colourist of all.

These engravings of tropical fish have been variously enlarged to become more or less the same size, then coloured with brilliant watercolour inks. It is important to use transparent paint or ink so that the patterns of the fine engravings show through the bright colours and are not obscured. The watercolours need mixing with diluted PVA: don't be alarmed by its milky appearance as it will dry clear, with a slight sheen. When the paint has dried, another coat of undiluted PVA strengthens and seals the paper.

When designing your frieze or shoal, it is a good idea to use small dots of Blu-Tack to stick the fish to the wall – you can move them around and overlap them until you are happy with the arrangement. They can then be removed one at a time and stuck up permanently with wallpaper paste.

YOU WILL NEED

Black-and-white photocopies of fish (Sheet 12)
Watercolour Inks in Cyclamen, Turquoise, Tangerine and April Green
Watercolour palette or several saucers
PVA
Artist's brushes to apply the colour
Scalpel or sharp scissors
Blu-Tack
Wallpaper paste
Pasting brush
Small decorating roller
Clear varnish
Soft cloth

1 Dilute the watercolour inks on a palette or in saucers with a 1:1 mixture of PVA and water, then have fun colouring the fish.

2 When the inks have dried, apply a covering coat of undiluted PVA to seal the colours and strengthen the paper. Use a scalpel or a pair of sharp scissors to cut the fish out carefully.

3 Plan the arrangement of your shoal and use small bits of Blu-Tack to stick the fish on the wall temporarily. Step back to judge the

effect. You may prefer to use the fish as a frieze all around the room. The Blu-Tack procedure will allow you to try all the available options before committing yourself to a final design.

4 Having made your decision, remove the fish one at a time and apply wallpaper paste to the back of each with the small roller or a brush. Paste each back in its place on the wall, overlapping some for the 'shoal' effect illustrated above.

5 Run the small roller over the pasted–up fish to force out any air bubbles and excess paste, then wipe the edges clean with a soft cloth. When dry, paint each fish with a coat of waterproofing, clear varnish.

CLASSICAL FIREPLACE

FIREPLACES CAME BACK into fashion on a wave of nostalgia in the late 1970s, having fallen from grace in the fifties and sixties, when 'modernity' was all the rage. Central heating was there to keep everyone warm and making fires was something country folk did with fallen trees. Eventually, people realized that a fireplace is far more than a heat source – it's a warm focal point, somewhere for guests to congregate and a very convenient place to display your Christmas cards and party invitations.

Fireplace design has not really changed much. The most popular styles are still Victorian or Edwardian, and plenty of reproductions, good and bad, are on the market. This is a project for people who recognize what the past has to offer in the way of rich source material. The photocopied fireplace borrows images from classical times, but shakes them up and rearranges them to make something bold, new and decidedly different.

The style suits any flat surface – wood, slate or marble. It is a good way to improve a plain, new surround or to lighten up a heavy old one. An old house may have marble or slate fireplaces that need time, skill and money to restore properly. If these commodities are in short supply, then photocopies will provide an instant lift without damaging the fabric or reducing the value of your relic.

When selecting images to photocopy, it is useful to make a scale drawing of the fireplace first. This can be used to plan and position the elements of the photocopied designs. The balance between the black and white areas needs to be considered, particularly on the upright sections, where there should be enough black to take the weight of the patterns above it. This is eclectic design, so isolate the basic shapes and play around with the scale of the enlargements. The last thing you want is for it to look symmetrical; it is important to be bold, so when you have made your decisions on the small scale drawing, have the enlargements made and go for it!

YOU WILL NEED

Tape measure
White emulsion (latex) paint
Tracing paper
Scalpel
Ruler
Photocopies of friezes and columns
Wallpaper paste
Scissors
Clear matt varnish
Brushes for paint, paste and varnish
Soft cloth

37

1 Measure all the structural elements of the fireplace surround, including the mantel shelf edge and the side edges which abut the wall. Make a reduced scale drawing of the whole fireplace on a manageable piece of tracing paper.

2 Select images of the right proportions to photocopy by laying the tracing-paper scale drawing over them. It is best to use a single enlarged panel above the fireplace and square motifs in the corners either side, but the other parts of the fireplace surround can be covered with a varied selection of cropped or complete blown-up images.

3 If you are unsure of the effect of juxtaposing different styles or images, make a photocopied scale plan of the fireplace with your designs in place, so that you can really judge what your design will look like, before moving on to enlargements.

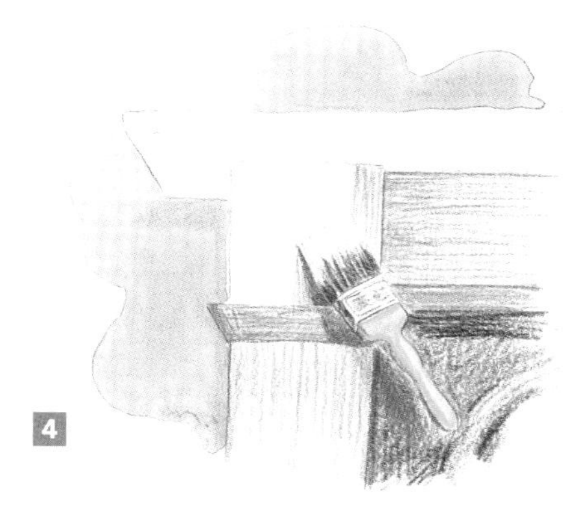

4 Paint the fireplace with two coats of white emulsion (latex) paint to give good coverage. Leave until bone dry, at least overnight.

5 Have enlargements of all the motifs made to the proportions of your fireplace. The enlargements are best made onto A3 paper, so that you have as few joins as possible. Cut them out with a scalpel to fit the shapes of the various elements of the fireplace.

6 Brush wallpaper paste onto the back of the photocopies and stick them onto the fireplace surround. If there are tricky corners or recesses, these are best cut with scissors when you are actually putting the paper up, as you would when wallpapering.

7 Smooth the paper with a clean, soft cloth to get rid of any air bubbles and to wipe off any excess paste. Leave to dry.

8 Finally, apply one or two coats of clear matt varnish all over the pasted-up photocopies, paying special attention to the edges and joins. This will seal and protect the fireplace.

CIRCUS FRIEZE

GIVING CHILDREN SOMETHING interesting but repetitive to look at when they lie in bed may have the same effect as counting sheep and lull them off to sleep. There is nothing sleepy about these striped 'circus tent' walls, though. They were created by measuring up the wall with a width of card, then running masking tape down the lines. A small roller was then used to fill in the stripes with bright red emulsion (latex) paint. It's a fun way to brighten up a dark room and make it seem like a friendly place.

The images used for the frieze all appear on Sheets 17, 18, 20 and 26, and they have been photocopied in black and white, so it is a very economical project. Many of the coloured papers that copy shops have on offer tend to be in pastel shades, because most of their business comes from printing documents that have to be legible. Ask if they have any brilliant colours – some do carry stock of very bright paper to use as chapter or section dividers. If not, buy some paper of your own in A3 size – but remember that it must be smooth textured to go through the machine.

When positioning the border, think of the child's height, not your own, and put it up where he or she can see it. The project can be tailored to various themes: it could use pictures of a child's favourite subjects – football, ballet, animals, trains, boats or planes, for example – and by using different coloured papers, the whole mood can be changed. Experiment a little; the cost is minimal, and the first plan may not always be the best one.

YOU WILL NEED

Tape measure
A3 white paper
Paper glue
Pink, green and yellow sheets of A3 paper
Sharp scissors
Scalpel
Cutting mat or piece of card
Spirit level
Pencil
Wallpaper paste
Brush for pasting
Small decorating roller

1 Measure the distance that you want your border to cover, halve it, then divide that by the length of an A3 sheet (42cm) to find out how many copies you need. For our project, this number should then be divided by three to find the number of sheets you need in each colour – pink, green and yellow.

2

2 Take a sheet of plain, white A3 paper and fold it in two lenthways. Unfold. Each half is now the width of the actual room border.

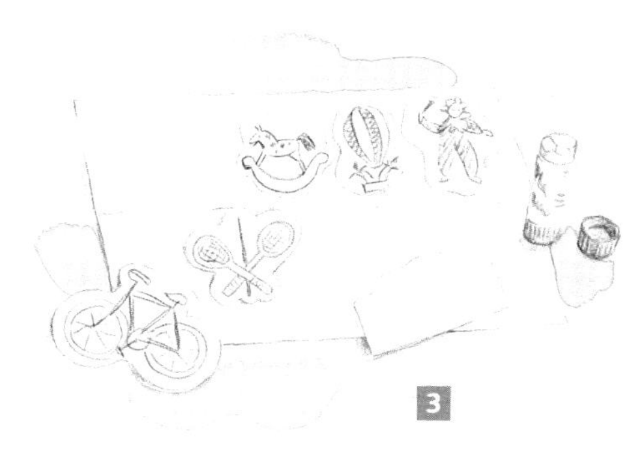

3 Enlarge the motifs (from the back of the book or your own choice) on the photocopier to a maximum of 4 inches (10 cm) in height. Trim the images with scissors, leaving some white around the edges, and arrange them along the width of the border. When the spacing looks good, stick them down using any make of paper glue.

4 Make a sheet of black A3 paper by printing with nothing to copy on the photocopier. Cut out three strips of black paper – two about ¼ in (6mm) wide and one about ½ in (12mm) wide. Stick the wider strip exactly on the centre of the crease in the middle of the white paper with stuck-down circus images and the two smaller ones parallel at each edge of the paper.

5 Ask the copy shop for the photocopied size to be reduced by 5%, to 95% (this allows a bit of leeway in still achieving the whole image, in case the paper doesn't feed through quite accurately each time). Make the required number of copies on A3 sheets of coloured paper, remembering that each sheet will be cut in half to make up the border.

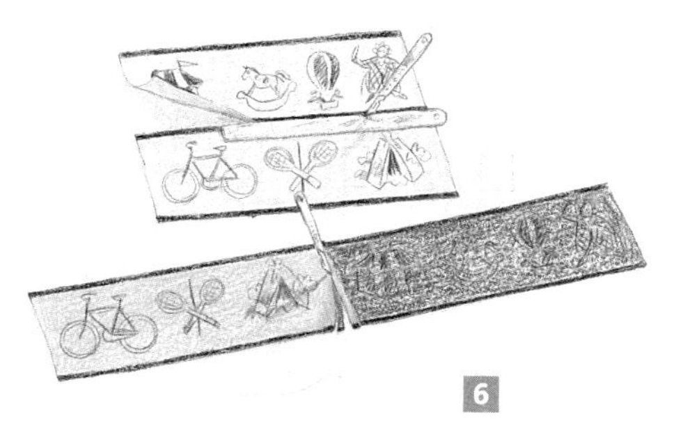

6 Cut the A3 sheets in half lengthways, using the scalpel on the cutting mat or card. Trim the short ends by overlapping adjoining sections slightly and cutting through both at once – this will ensure that they meet perfectly, without any gaps.

7 Using a spirit level to make sure that the line is straight, mark pencil dots around the room just below the top height required for the frieze. Brush the back of the frieze sections with wallpaper paste and stick them onto the walls, using the pencil marks as a guide. We arranged our frieze as alternating sections of pink, green and yellow. Smooth over the frieze with a small decorating roller.

TIMEPIECE PELMET

FASHION AFFECTS ALL areas of our lives and window treatments are certainly no exception. They are also a major expense, so it is good to discover ways of introducing style at a very low cost. If the main focus is a grand, theatrical pelmet, you can get away with using inexpensive fabric for the curtains – ticking, suit lining, calico or muslin all look good and are only a fraction of the cost of 'proper' curtain fabrics.

This pelmet took a morning to make, assemble and install. The most important part of the project is choosing the image for your photocopy enlargement: it needs to be long and thin, because the depth will increase in proportion to the width. The first thing to do is to measure the size of the window, including the frame, which will allow you to calculate the width of the pelmet. Then you can decide upon the depth, which will depend upon how much of the window you want to cover. Think about the overall proportions of the room, because the pelmet could be over-powering if it is too big and flamboyant.

Scale the measurements of the actual pelmet to bring the length down to about 6 inches (15 cm), then make a scale drawing of the shape on tracing paper. This can be placed over patterns and mouldings to see whether the proportions will fit the pelmet. It looks best to have the pelmet 'wrap around' the sides, so you need to choose patterns which will accommodate this area as well.

Pelmets are by their very nature 'over the top', but the designs don't have to be – it's up to you to find the style and the images that suit your home and just apply the essence of the idea to make something quite unique.

MAKING THE BOX

YOU WILL NEED

15 mm MDF (medium density fibreboard) for the top 'shelf' of the pelmet. A 6 inch depth (15 cm) should be enough for most windows. Measure your window to find the length, and calculate the required amount of MDF.

6 mm MDF – 6 inches (15 cm) or more deep, by the length of the 'shelf', plus half an inch (1.25 cm). This makes an allowance for the two side pieces, whose ends should be covered by the front panel. The side pieces measure 6 x 6 inches (15 x 15 cm) on our pelmet, but you should adjust the size to fit your shelf width and front depth.

Panel pins

Small hammer

Wood glue

2 shelf brackets

Screws and wall plugs

Screwdriver

1 Apply a coating of wood glue to the sides of the 'shelf' piece of MDF. Attach the side pieces, applying glue to the contact edges and adding strength by hammering in a few panel pins as well.

2 Apply a thin line of wood glue to the front edge of the shelf and on the edges of each end piece, then fit the front panel onto the shelf. Once again, use panel pins to strengthen the fixing.

3 Using the shelf brackets as a guide, mark the position for their screws at the outer inside edges of the pelmet box. The brackets will secure the pelmet either side of the window frame (see step 5, opposite).

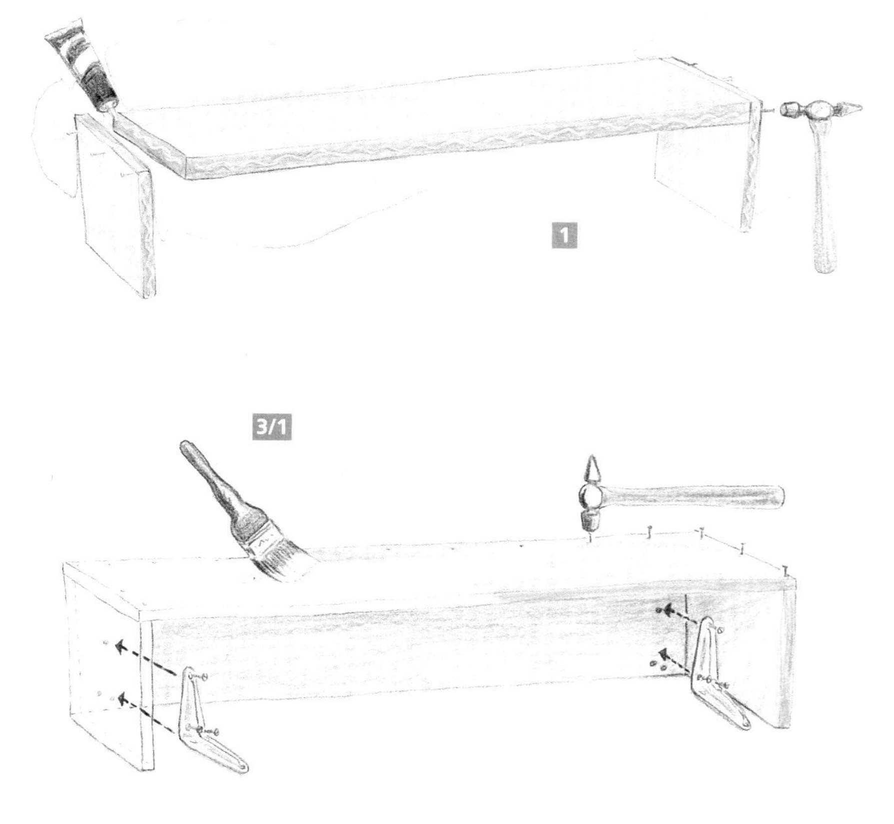

DECORATING THE PELMET

YOU WILL NEED

White acrylic primer
Paint brush
Scalpel and ruler
Cutting mat or card
Pencil
Tracing paper
Wallpaper paste and pasting brush
Selection of images photocopied and enlarged
Clear matt varnish and brush to apply it

1 Paint the pelmet box with white acrylic primer and allow to dry.

2 Select your images for the pelmet by overlaying the tracing paper scale drawing, as described in the project introduction above. Cut them out and paste them onto a sheet of A3 white paper, following the proportions of the scale drawing.

3 Enlarge the paste-up to create the number of sheets, with overlaps, needed to cover the area of the pelmet. Here, four A3 sheets of enlargement were needed. Overlap the 'tiling' areas of the photocopied enlargements and trim with a scalpel, along a ruler edge, onto a cutting mat or card, so that the edges butt up accurately, with no gaps.

4 Brush the outside area of the pelmet with wallpaper paste. Paste the back of the photocopied A3 sheets and stick them onto the pelmet, abutting the edges of the enlargements. Leave to dry.

5 Brush over with a coat of clear matt varnish and leave to dry. Drill and plug holes in the wall either side of the top of the window. Fix up the pelmet by screwing the brackets to the wall and then to the pelmet. Hang curtains for the finished effect.

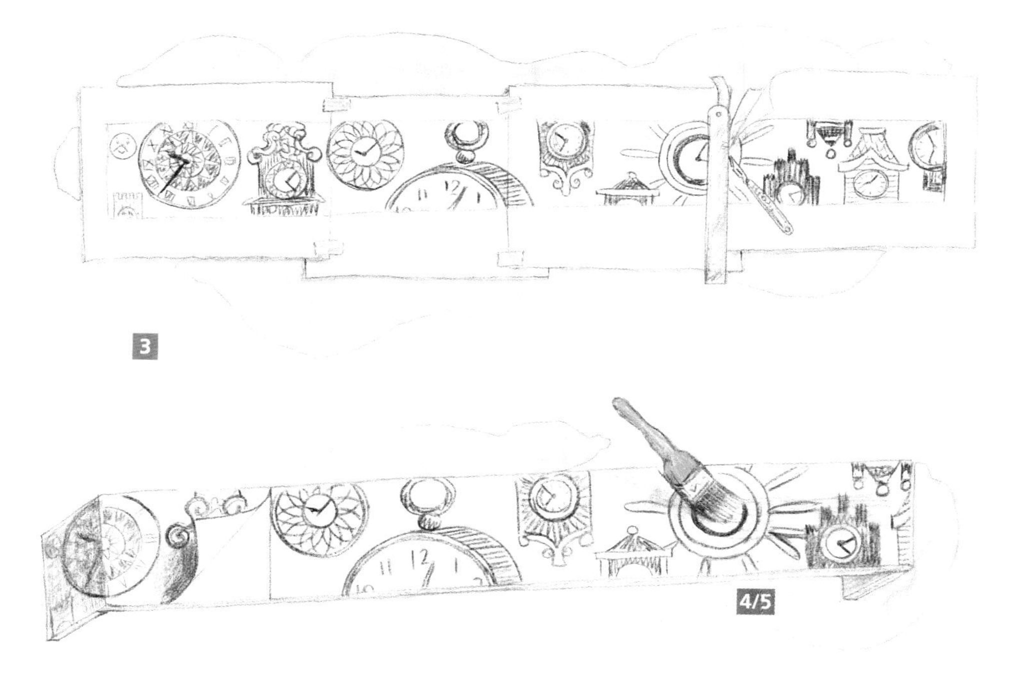

KITCHEN IMPLEMENTS *TROMPE L'OEIL*

OLD KITCHEN IMPLEMENTS are highly sought-after nowadays, but they are in limited supply and have become expensive antiques. Glossy homes magazines carry photographs of kitchen walls laden with quaint old colanders, ladles, spoons and grinders. The effect they produce is immensely stylish and comforting, but it costs the earth. Luckily, a photocopied *trompe l'oeil* arrangement which looks almost like the real thing provides an alternative.

The utensil illustration used for this kitchen wall appears in a very old book about ancient and modern furniture and woodwork. The utensils all feature in a woodcut taken 'from the original preserved in Rome'. The actual woodcut illustration in the book was small – only 3 inches long – but the quality of line was excellent, as can be seen from the crispness of the enlargements on our finished wall. The illustrations look surprisingly contemporary, but are all the more interesting because we know their age.

It is most economical to do all the enlarging on a black-and-white copier, then move on to a colour copier, choosing a colour to replace the black. The inspiration for the cream and green was a linen tea-towel. It's a good idea to base your colour scheme on something already in the room – your enamelled pans, a set of tins or your favourite tablecloth, for example. The wall paint colour should match the background colour of the photocopying paper as closely as possible. Once the shapes have been cut out and pasted up, they should appear to be outlines, not cut-out solid shapes, to achieve a successful *trompe-l'oeil* effect.

YOU WILL NEED

- **Kitchen implement images to photocopy (see Sheet 22)**
- **Cream emulsion (latex) paint - to match the background paper of the photocopies**
- **Blu-Tack**
- **Wallpaper paste and brush to apply it**
- **Scalpel or sharp scissors**
- **Small decorating roller**
- **Soft cloth or sponge**

1 Have enlargements made of the kitchen implement engravings on Sheet 22, or of others of your choice. Their size will depend upon the location – as a rough guide, the longest of ours measures 16 inches (40cm). Although the originals are black and white, these copies were done on a colour copier, substituting green for black and printing onto cream paper instead of white.

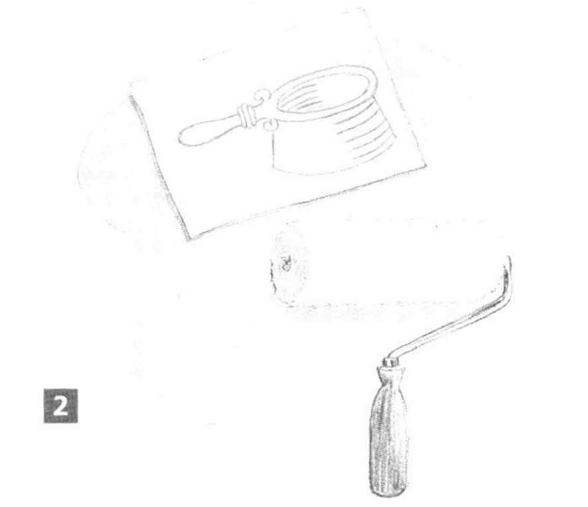

2 Paint the wall with cream emulsion (latex) and allow to dry. Try to get a very close match to the cream paper: taking a sheet of the paper along to the paint supplier is a good plan, as he can help to mix a matching paint.

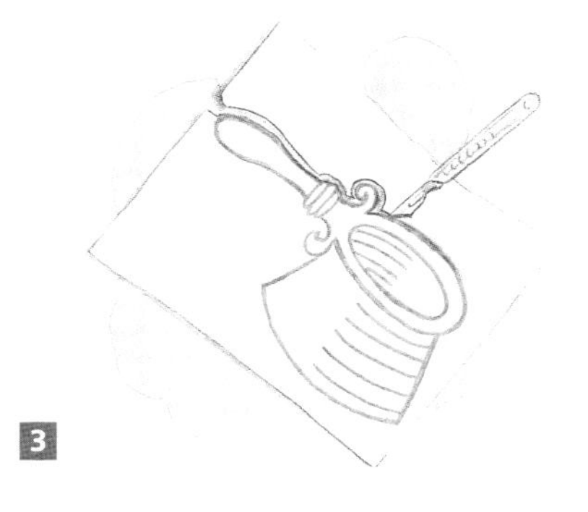

3 Cut out the kitchen implement photocopies, leaving a round-edged border of cream paper around each rather than cutting in too close to the images. Use Blu-Tack to make a temporary arrangement of the shapes on the wall, which will allow you to step back and judge the effect.

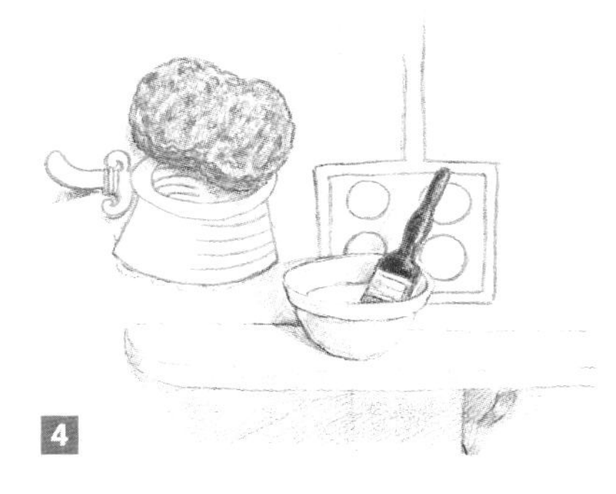

4 Apply a coat of wallpaper paste to the back of each implement and then stick them all into position. Smooth out any air bubbles using a small roller and wipe any excess paste from around the edges with a soft cloth or sponge.

Chapter Three

Fabrics

DURING THE PAST decade, more people than ever before have begun to decorate their homes in a very hands-on way. Previous generations introduced pattern and texture with wallpapers and fabrics, but now, thanks to the masses of instructional books, magazines and television programmes on the subject, most of us feel confident in applying pattern ourselves using stencils, stamps or paint finishes.

However, one area is still largely the preserve of professional designers – textile printing. There are many reasons for this, not least that fabric and the special wash-proof inks needed for the job are more of an investment than a pot of paint, and if mistakes are made, they cost more to correct. Decorating fabric is not yet perceived as an area suitable for amateur enthusiasts. This is about to change, and soon everyone will be painting and printing their curtain and cushion fabrics, just as they now do their walls and furniture. Using photocopies to print onto fabrics is an easy technique, which will give you the courage to create your own unique fabrics for the home, and some to wear as well.

There are two avenues to follow: one is the copy-shop print onto fabric, which is done on a heat press, and the other is a do-it-yourself method, a lot more time consuming but much, much less expensive. A gel is applied to the surface of a photocopy and the print is then smoothed onto a fabric. Once the gel has dried, the fabric is saturated with water to remove the paper and the image is left behind. Ideally, this is done on small areas, because the soaking and rubbing off of paper is better confined to a bowl than a bath. One way of getting around this is working the images into a bigger patchwork pattern, as we did with the vegetable printed curtain on page 60.

The other method only requires the selection of an image and a suitable fabric and a trip to a copy shop that advertises T-shirt printing. It is done using a thermal process: a special transfer paper for the photocopy, then a heat press to bond the print to the fabric. Provided that you choose a white fabric with a very smooth surface texture, the results are as good as printing onto paper.

When we first suggested printing onto fabric other than T-shirts, the operators in our local copy shop looked dubious, but were willing to try. We chose well-washed and ironed cotton sheeting, and the results were staggeringly good. The main drawback is, unfortunately, the cost, which is relatively high at about five times that of an ordinary colour photocopy, excluding the price of the fabric itself. However, the cost of making a set of unique cushions, in the design of which you can give free reign to your imagination, will still not be more than buying them ready made.

Inspiration for designs can be found all around you; sometimes, the mundane and obvious just needs looking at from a different angle. Use your camera to take close-up photographs of things around the house: for example, the cat's fur, the inside of your cutlery drawer, a plate of breakfast cereal or noodles, a bundle of coloured ribbons, the contents of a jewellery box or a row of toothbrushes. Once processed, enlarged, photocopied and transferred, any one of these could become a fabric that would make the style press sit up and take notice.

COOK'S APRON

THERE IS USUALLY someone in every family who is renowned for making the best-ever version of a popular dish. It could be Uncle Arthur's egg and chips, your cousin Daisy's chocolate fudge brownies or Grandma's apple pie. Compliments are always welcome, and if you are lucky may achieve the desired result – securing you a lifetime's supply of your favourite treat. But present the cook with a personalized apron that declares her supreme, and culinary results are practically guaranteed!

You can buy a plain, cotton apron from a catering supplier or make one yourself, using an old apron as a pattern. The image that you use will depend upon the dish that is to be featured. We found this lovely drawing of a pie in an old cookery book bought in a junk shop and would suggest that you pursue a similar route. The 'No 1' motif appears on Sheet 9 at the back of the book; simply photocopy it and enlarge your chosen 'dish' to a suitable scale. (A number of food motifs which may be suitable appear on Sheets 8 and 9.) If you want to make your gift even more personal, you could use rub-down transfer lettering in a similar italic style to add the cook's name to your design. Lettering can also be designed on a computer, printed out, then photocopied.

To transfer the design onto the apron, either follow the instructions for the do-it-yourself method described for the patchwork kitchen curtains on page 62, or have the image heat-transferred at your local copy shop.

MAKING YOUR OWN APRON

You need a little over half a yard (about half a metre) of strong cotton fabric with a smooth surface texture, such as cotton duck. Using an old apron as a pattern, draw out the basic shape. Turn back a half-inch (1.25 cm) seam, using a zig-zag stitch and snipping the fabric on the curves so that it folds back flat. Now turn the seam again and over-stitch all around the edge. Attach tapes at the top of the apron (the adjustable rings are entirely optional) and sides.

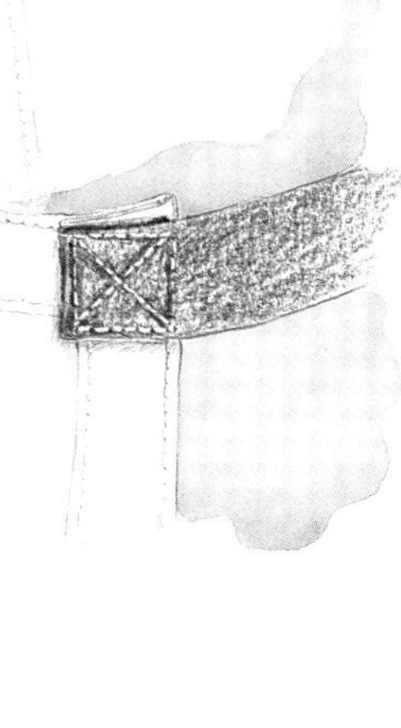

№ 1

DESIGNING THE TRANSFERRED IMAGE

1 A lot will depend upon the picture image that you choose, but you can judge the proportion of the lettering to the image by looking at our example. You want the design to convey your message directly, so make sure that it is big enough to do so.

2 The joy of black-and-white photocopying is that it is very cheap, so make several copies of the images and use them to plan out the design. Cut out the lettering and the illustration, fold each in half to find the centre, then open them out again, leaving the centre clearly marked.

3 Now you can move the images closer together and further apart, keeping them centred, to find the most comfortable spacing. As a rule, they will look right if the space between them is roughly half the height of the lettering – a larger space will separate them too much and prevent them from reading as a single message.

4 If you are adding the cook's name, use lettering about half the height of the 'No 1', and apply the same spacing and centring rule.

5 When you are happy with your design, stick the elements in position on a sheet of paper and have it photocopied. Remember that if you want to use the do-it-yourself method with the transfer gel, you should have all the lettering reversed on a photocopier so that it reads correctly when transferred.

SAILOR'S TOTE BAG

YOU NEED SO many bits and pieces when you go out in the sun these days – sunblock, a spare T-shirt, a purse, flip-flops, something to read, your sunglasses. Life is pretty complicated, even for the seven-year-old we catered for here.

The answer, of course, is to use a bag that's recognizably your own; and what better way to identify it than by having your snapshot printed onto the front? Otherwise, you could make a themed bag to suit any hobby or interest, from ballet to bungee jumping. Just choose a favourite photograph and take it down to the photocopy shop.

This photocopy has been taken from a holiday snap, and enlarged up to A4 size on a colour copier. The photocopier operator then transferred this print onto smooth, white cotton fabric, using the heat transfer method. After that it was plain sailing – or sewing, actually – to run up a few seams on the sewing machine and insert eyelets and rope to make the tote bag. These are quite appropriate for our nautical theme, but a simpler version of the tote bag can also be made by just threading a cord through the top seam to make a drawstring.

YOU WILL NEED

Just over a yard (about one metre) of smooth, white cotton fabric – a washed-out cotton sheet is perfect
About three-quarters of a yard (80 cm) of white petersham binding
Scissors
Snapshot of yourself
Pack of large eyelets
Small hammer
Soft cotton rope to fit the eyelets
White thread
Sewing-machine (or a needle and patience)

1 Cut out the two pieces of the white cotton fabric for the tote bag, each measuring about 14 x 19 inches (35 x 50 cm).

2 Have an A4-size colour copy made of your selected photograph and ask the copy shop to heat transfer the photocopy onto one of the two pieces of cut fabric. The print will look best positioned about 3 inches (7.5 cm) from the bottom of the bag, centred between the sides.

3

3 Place the two pieces of fabric one on top of the other, with the printed side uppermost. Sew a seam about half an inch (1.25 cm) in around the side edges and bottom edge.

4

4 Now turn the resulting bag inside out and sew the same seam width again, tucking the raw edges inside. This is called a French seam, and it has the advantage of strengthening the seam and preventing the fabric from fraying.

5 Line the strip of petersham binding up with the top edge, about half an inch (1.25 cm) down from the edge. Turn over the half inch and stitch it to the petersham.

6

6 Turn the petersham over and sew to the bag with a zig-zag stitch along its bottom edge. The binding will give a good, stiff seam to hold the eyelets.

7

7 Use a small hammer to insert the eyelets at regular intervals around the top of the bag. Eyelets are sold in sealed packs which include an assembling tool and instructions. The eyelet and washer are placed in the tool and slid one over and one under the fabric, then banged in with a hammer.

8 Thread the soft rope through the eyelets and tie the two ends together with a simple knot.

8

PATCHWORK KITCHEN CURTAINS

THE PROCESS INVOLVED in transferring a photocopied image onto fabric is quite time consuming and requires elbow grease. There is something magical about it, nonetheless, especially when you realize that the engraved images of vegetables used in this project originally appeared as tiny illustrations in a turn-of-the-century encyclopaedia. One of the drawbacks of the process is that the transfer prints need to be sealed after they have been applied to the fabric, which leaves a shiny, plastic-feeling surface. However, this is fine for kitchen curtains, especially if you use a glazed cotton for the contrasting squares to complement, rather than fight against, the transfer finish.

A sewing-machine makes light work of patchwork squares, especially large ones like these, and it allows you to give the transfer prints maximum impact without the job becoming too laborious. The measurements used here are not arbitrary, but are designed for curtains to fit a window 45 x 56½ inches (114 x 142 cm). Just adjust the size, number and arrangement of the squares to fit your own window.

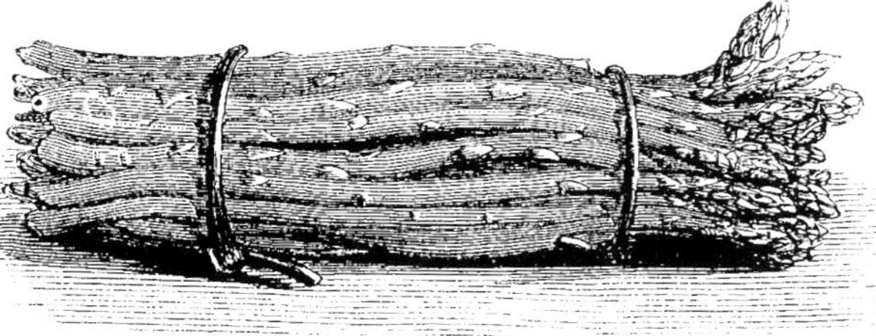

YOU WILL NEED

- 1 white cotton scrip to use for a test
- 14 white cotton squares measuring 8 x 8 inches (20 x 20 cm)
- 4 black cotton squares measuring 8 x 8 inches
- 12 green cotton squares measuring 8 x 8 inches
- 8 green cotton squares measuring 6 x 6inches (15 x15 cm)
- 4 strips of black cotton measuring 23 x 6 inches (57 x 15 cm)
- 4 strips of black cotton measuring 38 x 6 inches (95 x 15 cm)
- 4 yards (3 metres) striped green cotton lining fabric
- Curtain header tape – the iron–on type used here (buy 2 yards [1.83 metres] and use slightly less)
- 2 tubes of Image Maker manufactured by Dylon
- Large bowl
- Medium-sized, square-tipped artist's brush
- Sponge
- Photocopies (see Sheet 8) to fill the white squares plus extra one or two for testing

1 Experiment with different images, enlarging or reducing them on the photocopier. The process used for changing a photocopy into a transfer will reverse the original image. If you want it to appear the same way around, you should have the image reversed on the photocopier before making the prints. Remember, only the line, not the background, will print. Look for images with good line quality; old engravings reproduce well. Make five copies of each one.

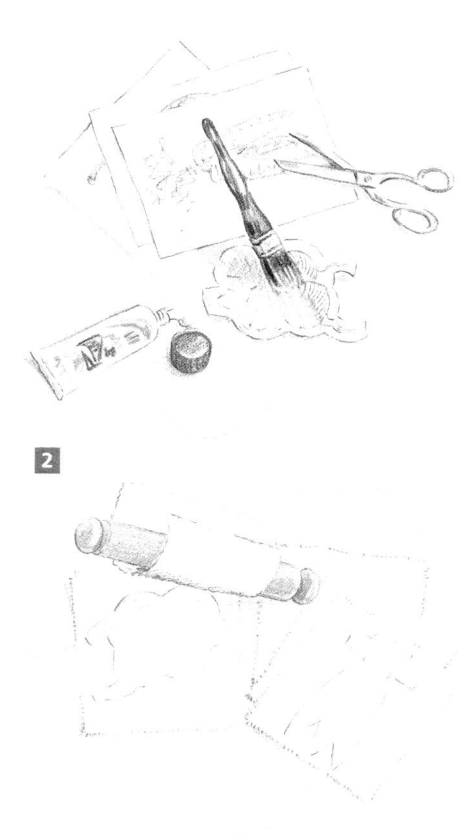

2 Cut around the photocopied images, then apply a thick coating of Image Maker to one of them. Place it centrally on top of your test scrap of material. Smooth the back to ensure that all areas are in contact. Repeat the process, putting the various photocopies onto all the white squares, and leave them overnight.

3 Fill a bowl with cold water and float the white fabric squares in it until totally saturated. The paper looks white until it absorbs the water, so wait for any white spots to disappear. Remove the test print from the water and place it on an absorbent surface, like an old towel. Use the sponge to rub off all the paper. If it is wet enough, it should lift clear in one piece, but it is more likely to disintegrate and need firmer rubbing in some places. The transfer is quite tough, but it is possible to lift it from the fabric if you are not careful. You need to complete the process to fully understand it, so the test print is essential.

4 Remove the paper from all the squares and leave them to dry. If some powdery remains of paper appear as the fabric dries, wet it again and rub with the sponge. If you do this 'dry', the print will lift.

5 Seal the print with a thin finishing coat of Image Maker. It dries to a sheen, so treat the whole square, not just the transferred image.

TO MAKE UP THE PATCHWORK CURTAINS

1 Arrange the squares on a flat surface, moving the different images until you are happy with the balance, then join the squares together in strips, allowing roughly half an inch (1.25 cm) for the seams. Sew the strips together to make the central panels of patchwork.

2 Sew corner squares of the curtain to the side strips, then sew the whole of each to the sides of the patchwork.

3 Sew the top and bottom strips of the curtain to the main panel, then to the corner pieces. This part can be tricky, so it is advisable to pin and tack (baste) the adjoining edges before machining. Press all seams flat on the wrong side.

4 Place the lining fabric face-to-face on top of the patchwork and cut it to size, leaving an additional seam allowance at the top edge to turn inside.

5 Machine around the side and bottom edges, then turn the lined curtain inside out (like a pillow slip). Press seams flat on the lining side.

6 Turn a seam inside using the top edge of the patchwork and the lining, and machine along the outside edge in matching thread.

7 If you are using iron-on curtain heading tape, cut it to size and follow the manufacturer's instructions to attach it; otherwise, machine the tape in place with matching thread.

FRANKED T-SHIRT

We all know the saying 'been there, done that, bought the T-shirt'. Although we may cringe, we still scour the clothes rails in foreign lands, hoping to find a stunningly tasteful or totally over-the-top tacky reminder of our time away. This project gives you a chance to make your own souvenir to tell everyone exactly where and when you were languishing on a beach or out on the town.

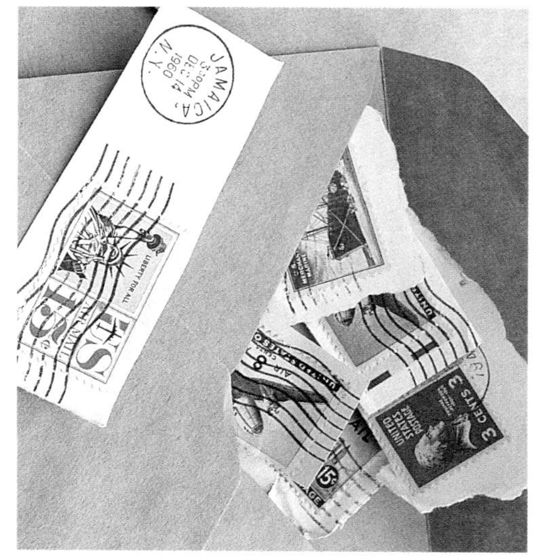

Your first stop after checking into your hotel should be the local post office. Here you buy a stamp (ideally a pretty one) and send yourself a letter or postcard at your holiday address. Who knows what the postal system is like, so do this on day one! When you receive your mail, you will have a stamp with the name of the country you are in, which will have been franked with the name of the postal district and the date. When you get back home, take your envelope or postcard down to the local colour print shop and have an enlarged print made to fit the front of a T-shirt. You can either get the shop to use the heat transfer process to put the photocopy onto the T-shirt or have a reverse print done and do the job yourself with Image Transfer Gel, as outlined below. The result will be an exclusive, one-off, souvenir T-shirt.

YOU WILL NEED

Franked postcard or envelope
T-shirt – white or a suitably pale colour
Image Maker , manufactured by Dylon
Brush
Ruler and set-square (triangle)
Sponge

1

1 Select the section of the stamped and franked envelope that you want to appear on the T-shirt. We chose just the stamp and frank mark, but you could include your name and holiday address as well.

2 Have the image enlarged on a colour copier so that it reaches the edges of an A4 sheet. Explain how you intend to use it and the print shop will make sure that you get the size right.

3 The gel will turn the print into a transfer, so you need to reverse the image on the photocopier first, otherwise it will all come out backwards. Ask for a mirror-image print.

4 Brush an even coating of Image Maker onto the print and position it on the front of the T-shirt. You should be able to judge this by eye, but use a set-square and ruler if it helps.

4

5 Leave the T-shirt for at least four hours – overnight if possible – then soak it in a bowl of cold water until thoroughly saturated. If you leave it long enough, you will be able to lift the paper away in one piece, leaving the print on the fabric. If the paper breaks up, just rub the print with a wet sponge until all the paper

has been removed. The T-shirt should be hung up to dry naturally; don't tumble or iron it dry.

5

6 Apply a sealing coat of Image Maker, according to the manufacturer's instructions. The print will wash well, but direct heat from a radiator or iron should be avoided.

CUSHIONS

IMAGINE THAT YOU could take a photograph of whatever you choose and then make it into a cushion cover. It may sound a bit of a strange idea, but take a look at this pair of cushions, with their seaside theme, and try and think of anything more perfect to soften your beach-hut bench or folding chair.

These photographs were taken using the close-up setting on an auto-focus camera, developed in an hour and then whizzed down to the copy shop. We cut the colour photographs down to give them a square format and asked for them to be enlarged onto A4-size paper. These in turn were trimmed into squares and heat transferred onto squares of cotton sheeting. The quality of the results is really fantastic: the cushions didn't lose any of the clarity of the photographic prints. This has a lot to do with the fabric that we used – a very smooth, pre-washed and ironed cotton sheet. The smoother the surface, the better the print.

Within three hours of the original photographs being taken, the cushions covers had been stitched together and filled with cushion pads. You could use this idea to make cushions for special occasions – Christmas cushions with holly wreaths would be good, or posies of flowers in the springtime – or simply to make an arrangement of colours and textures that match the rest of your room.

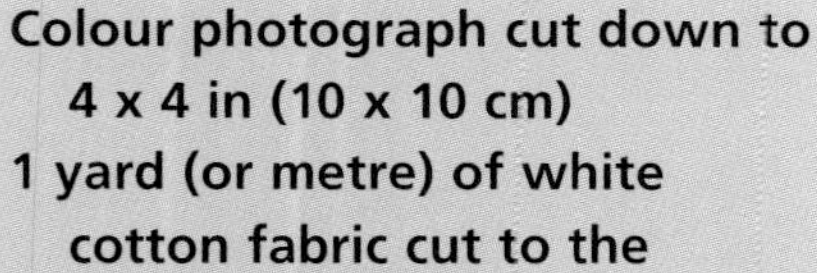

YOU WILL NEED

- Colour photograph cut down to 4 x 4 in (10 x 10 cm)
- 1 yard (or metre) of white cotton fabric cut to the following shapes::
  - 1 measuring 19 x 19 in (48 x 48 cm)
  - 2 measuring 19 x 12 in (48 x 30 cm)
- 1 yard (metre) piped bias binding
- 1 x 18 inch (45 cm) square cushion pad
- Pins and white sewing thread
- Scissors
- Sewing-machine with a zipper foot

1 Press the square piece of cotton so that its surface is really smooth. Take it, together with the colour photograph, to the copy shop and ask them to colour print the picture onto A4 paper and heat transfer the photocopied image onto the centre of the cotton square.

2 Lay the square, print upwards, on a flat surface. Lay the piped bias binding out around the cushion to judge the exact length needed, then pull back the binding by about an inch (2.5 cm) on both ends and cut away half an inch (1.25 cm) of piping. This will give a flat end of binding that can be sewn into the seam, rather than two bulky piped pieces.

3 Pin the binding around the edge, starting half way along one side, with the rounded side facing inwards and the flat edge lined up with the edge of the fabric. In order to turn the corners neatly, snip the flat part of the binding up to the stitching that holds the piping. Machine around the edge, using the zipper foot to get up close to the piping.

4 Sew a hem along one long edge of each of the other two pieces of cotton fabric. Lay the piped square print side up and line the unsewn long edge of one of the two smaller pieces up with an edge of the square. Make sure that the turned edge of the seam sewn on the smaller piece faces upwards. Now lay the other piece along the opposite edge to it, so that they overlap in the middle section.

5 Sew all the way around the outer edge of the cushion cover. You may need to adjust your machine to sew through the four layers where they overlap.

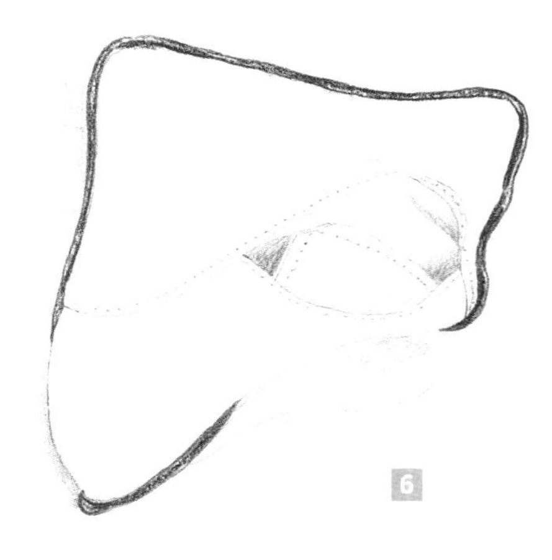

6 Trim the outer seam, nicking the corners of the cushion cover, then turn the cover inside out and fit the cushion into it. The overlapping small pieces of fabric cover the pad from two directions, giving a very neat finish, without the need for a zip.

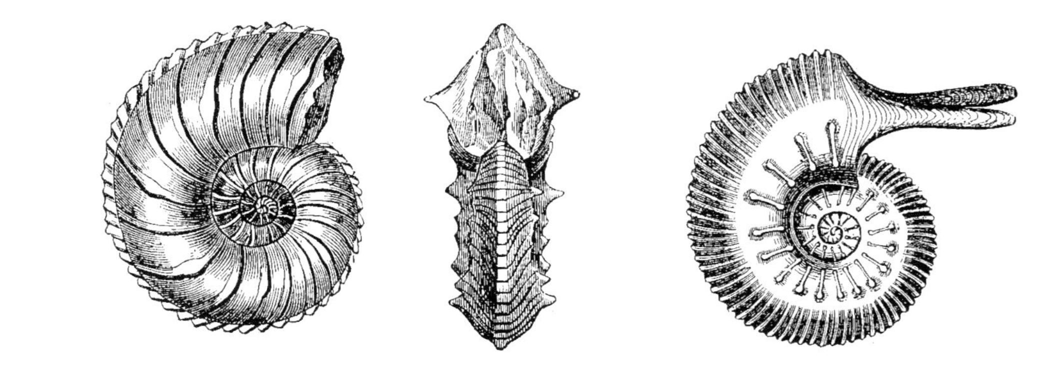

Chapter Four

Furniture

THE ART OF APPLYING paper patterns and designs to furniture is known as *découpage*, which comes from the French word meaning 'to cut out'. The Italian version of the technique, *l'arte del povero,* translates as 'poor man's art', which lends a welcome touch of accessibility to the subject. Historically, the process stems from the fashion for highly decorated, inlaid furniture made by Venetian craftsmen in the eighteenth century. The demand exceeded the supply and the cost was prohibitive. *Découpage* provided a means to mass-produce the best designs of the leading exponents. The original patterns were printed on paper, skilfully cut out, then pasted onto furniture, to be coated with many layers of varnish until the applied pattern was indistinguishable from an inlaid one.

Photocopying is ideal for this type of *découpage* and also for the more contemporary, textured effects we tackle in some projects. Photocopied images can be used to provide information, as we do with the medicine cabinet on page 82; to create an illusion, as with the turf table-top on page 76; or purely for the bright, colourful fun

of the effects. The photocopies do not have to be the main focus, either: they can be used in combination with other decorative crafts on the same piece of furniture. A wooden cupboard or chest could feature coloured, photocopied prints painted with antiquing varnish, surrounded by applied gilded mouldings, on a paint-finished background. The child's cupboard is a good example of the way that a number of techniques can combine to make something uniquely personal, but not necessarily traditional.

Acrylic varnish is easy to use, has no smell, dries quickly and the brushes used to apply it can be washed under a tap. Each coat that you apply will improve the look and the longevity of the piece of furniture, and you can use the varnishing process to add colour by tinting with acrylics or watercolours or for artificially ageing the piece by using antiquing varnish. This works best for rustic or folk art furniture, when you can't wait the fifty-odd years that it would take to develop a natural patina. Acrylic varnish comes in glossy, satin (a subtle sheen) or matt finishes and each suits different applications. The glossy is good for a bright, colourful, enamelled look, such as we used for the fridge magnets on page 102. The sheen of satin varnish adds a depth and touch of class, and the matt version provides an invisible protective coating that will not alter the quality of the image.

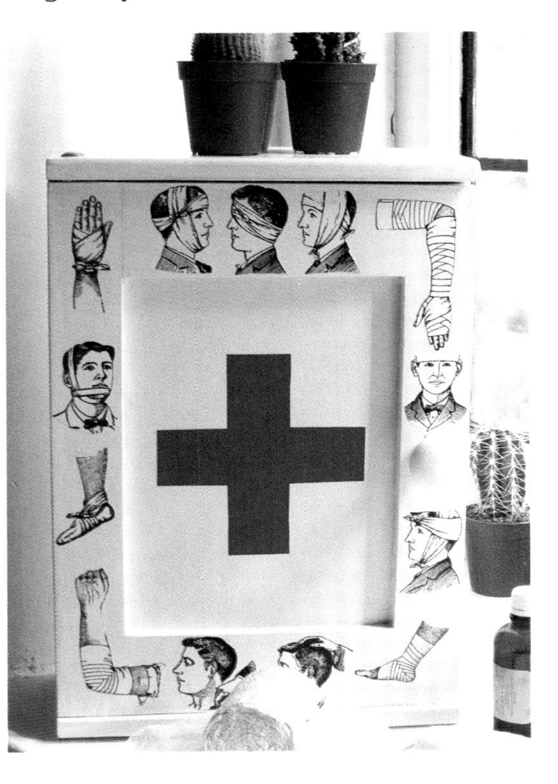

The projects in this chapter differ enormously, their only connection being that photocopies have been pasted onto moveable objects. We have tried to feature as many different approaches as possible in just five projects and hope that you will find the concepts themselves as inspiring as the actual finished objects.

CHILD'S CUPBOARD

ARTISTS LONG TO express themselves with the freedom and vigour of the very young, but sadly most are held back by the conventions learnt in later years. A three-year-old will grab the brush, soak it with colour and daub away, using a whole arm to make the movements. Small children appear at the play-school door bursting with pride and generously present you with the dripping image of the day. Give their confidence a boost by framing the best paintings, or use them to decorate a panelled cupboard such as this. If you have four paintings with a similar feel, then enlarge each one and cut it to fit the panel; otherwise, you could divide a single painting into four and enlarge each section to fit each of the four panels. This will give the effect of one large picture viewed through the panes of a window.

Children don't usually get excited about furniture, but this is no ordinary cupboard, so expect a positive reaction. The panels here provide a showcase for their art and the rest of the cupboard is converted into a 3-D blackboard with a coat of special matt black paint. All you need to provide is chalk and a duster.

The easiest cupboard to use would be one with existing panels, but you could customize plain doors by tacking on custom-made wooden mouldings to simulate panels. If you are doing this, stick the photocopies down first, then cover the edges with the mouldings.

Blackboard paint is available in both acrylic- and oil-based formulas that provide very good matt coverage, ideal for holding chalk lines. It will eventually develop a shiny surface from the chalky rubbing-out, but when this happens you can just apply a fresh coat of blackboard paint. The paint used on the top of the cupboard is gloss – a bright and practical option in a young child's room.

YOU WILL NEED

Small cupboard with panelled doors
Medium grade sandpaper
Brightly coloured gloss paint
Blackboard paint
White acrylic primer
Brushes
4 bold paintings done by a young child
Wallpaper paste
Quadrant beading - measure around the edge of all the panels to calculate the length of beading needed
Mitring block
Small saw
Panel pins
Small hammer
Wood glue
Clear matt varnish
Soft cloth

1 Prepare the cupboard for painting by sandpapering down any old paint or irregularities in the wood. Paint the top of the cupboard with gloss paint in the colour of your choice. Paint the areas inside the panels with white acrylic primer. Paint the rest of the outside of the cupboard with blackboard paint.

2 Make colour copy enlargements of the paintings to fit the dimensions of the panels. If you select the boldest sections of the paintings for the panels, the effect could be quite abstract.

3 Apply wallpaper paste to the back of the prints and to the panels and stick the photocopies in place, using a clean, soft cloth to smooth out any air bubbles and wipe off excess paste. When bone dry, apply a coat or two of clear matt varnish.

4 Measure each panel, then mitre quadrant beading on the block, cutting with a small saw, to fit inside them. Use wood glue and then panel pins and a small hammer to fix the beading in place. This will give a neat edge and prevent the photocopies from lifting when curious little fingers try to remove them.

LAWN TABLE

Breakfast can become a permanent picnic by covering your kitchen table with bright green grass. The first stage of this project involves taking a close-up photograph of a lawn, which can result in some funny looks if you do so in a public place! Get in as close as you can, keeping the focus really sharp, and take several shots to be certain of getting a good image to print from. Have the film processed in the usual way. Measure your table-top to find the number of photocopy prints needed to cover it completely and have the colour enlargements made to fit.

This table, found in a junk shop, has a formica top which needed roughening up with coarse sandpaper to provide the glue with a good surface to 'key' into. If you are using a wooden-topped table, it is a good idea to sand it down anyway to get rid of any rough edges that could tear the paper. The prints can then be cut out and matched together, and pasted down with their edges overlapping the edge of the table. Once the glue has set, you can trim the paper up to the edge and apply a coat or two of varnish.

Naturally, you are not limited to using grass on table-tops – you may prefer to photograph beach sand, a flower bed, clouds reflected in water or a plate of baked beans – whatever will bring a smile at breakfast time.

YOU WILL NEED

Suitable table
Sandpaper
Colour photograph of a lawn or other textured surface
Colour A3 photocopies of the photograph
Scalpel
Wallpaper paste
Brushes
Clear varnish
Soft cloth or sponge
Trim for the table edge (optional)

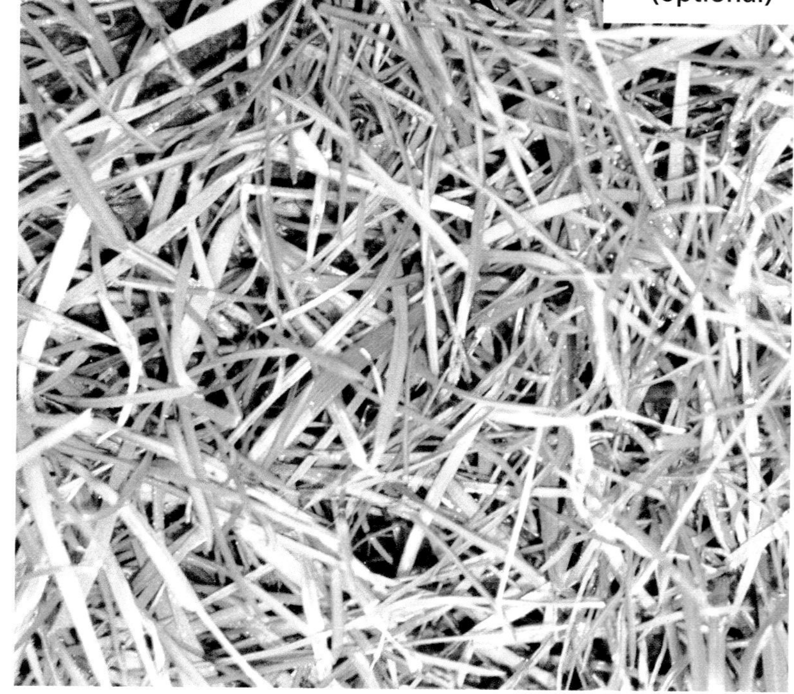

1 Choose the colour print with the sharpest focus to photocopy, because any blurring will be exaggerated in the enlargement. When you have the copies made, you can ask for the balance of the colours to be altered – the green has been made more dominant here. Trim the white edges from your prints with a scalpel.

2 Sandpaper the table-top and wipe away the resulting dust with a damp cloth.

3 Mix up the wallpaper paste following the manufacturer's instructions, then paste both the table-top and the back of each print as you go along.

4 Arrange the prints so that the edges butt up to each other without gaps or overlaps. Smooth any bubbles and blobs of paste out with a soft cloth or sponge, working towards the edges of the table and wiping off excess paste. Leave to dry.

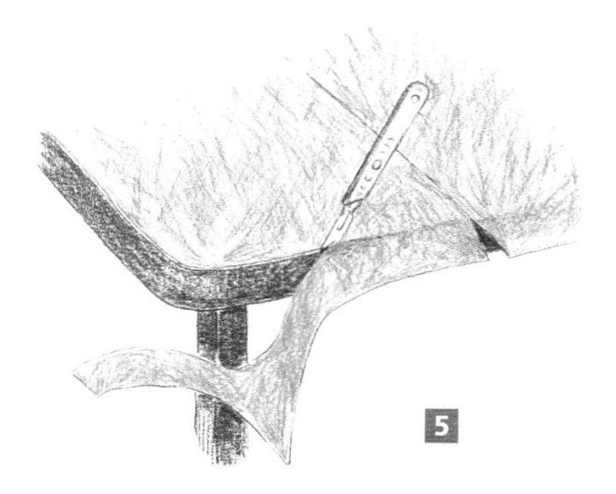

5 Use a very sharp scalpel to trim the paper right up to the table edge, then apply at least three coats of clear varnish, allowing the manufacturer's recommended drying time between coats.

6 An edging strip for the table can be bought from the kitchen department of a hardware store or from shops that sell car accessories. The strips are made of self-adhesive coloured plastic or screw-in rigid chrome, which would be perfect for a fifties diner look.

DECORATED CHAIR

A HUMBLE KITCHEN CHAIR may harbour delusions of grandeur that are crying out to be satisfied. Photocopied *découpage* can do the trick, lifting it from mundane to marvellous. This chair has been converted into a pastiche of the elegant French Empire style, using stark black and yellow and a combination of rosettes, pillars and decorative mouldings. These motifs were photocopied from a furniture design source book of the period on a black-and-white machine, then enlarged onto bright yellow paper. (Some chair and other furniture patterns are provided on Sheet 25.)

As this is a project about decorating existing chairs, of variable sizes and shapes, it is best to give general, rather than specific, instructions. Choose a wooden chair of similar proportions to ours and begin by measuring the broad back-rest. Say, for example, the size is 20 x 4 in (50 x 10 cm); the next step is to draw a scaled-down version of the shape on tracing paper (5 x 1 in or 12.5 x 2.5 cm). The tracing can then be placed on top of illustrations of various different mouldings, drawings, carvings, etc., to find something which is the right shape for the chair back. On this chair we used three separate pieces – two rosettes in square boxes and a centrally positioned moulding pattern. The length of the columns can be adjusted by cutting out or adding to their mid-sections.

The chair was first given two coats of emulsion (latex) paint, because the matt finish provides a good 'key' for the paste. The effect of the columns on a black background is slightly skeletal, especially if the chair is then placed against a dark background. This effect can be used to great advantage to slim down chunky chair legs and make them look more elegant.

YOU WILL NEED

Wooden kitchen chair with flat surfaces
Tracing paper
Pencil
Ruler
A3-size coloured paper
Images to photocopy
Sandpaper
Black emulsion (latex) paint
Brushes
Wallpaper paste
Soft cloth
Clear varnish

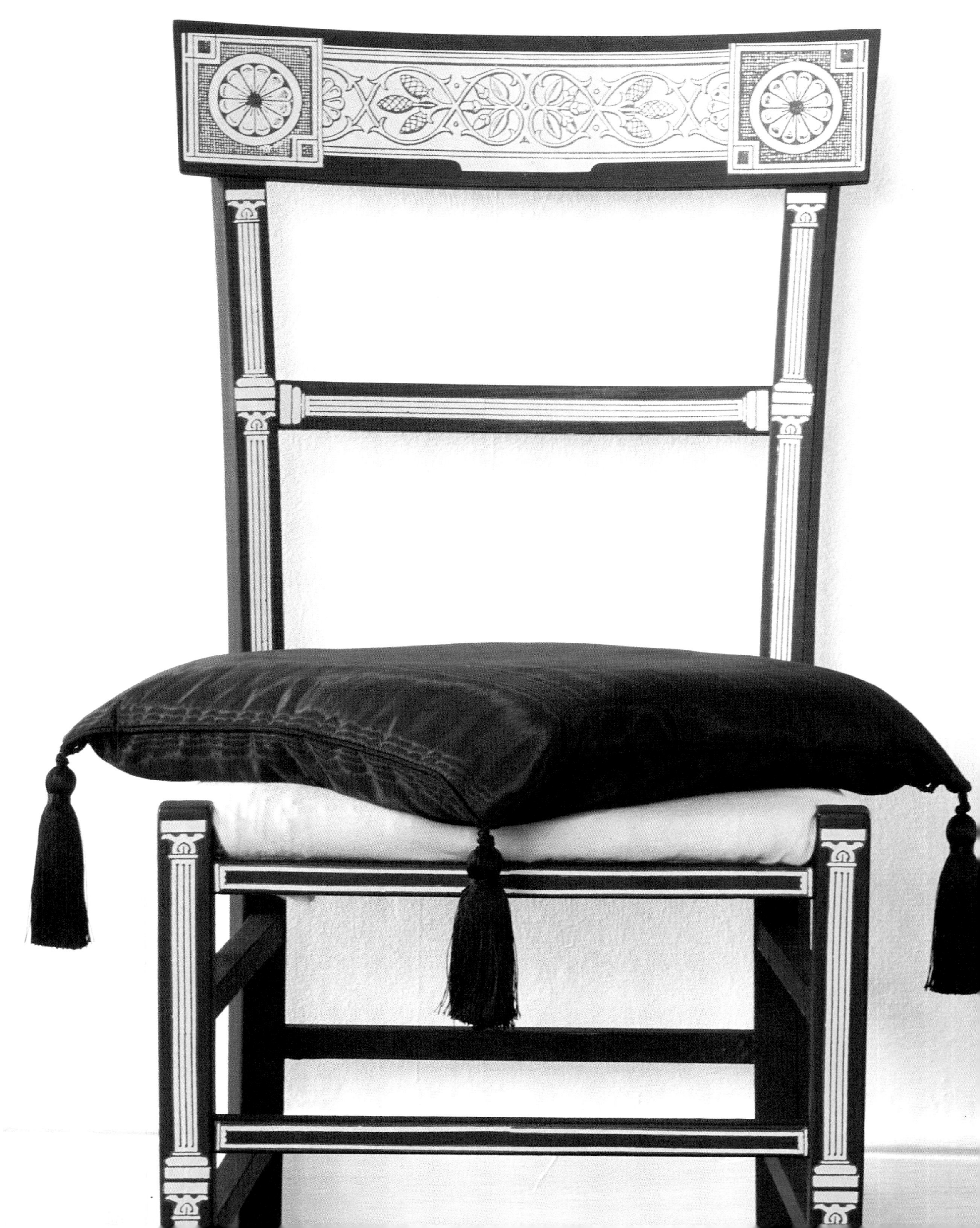

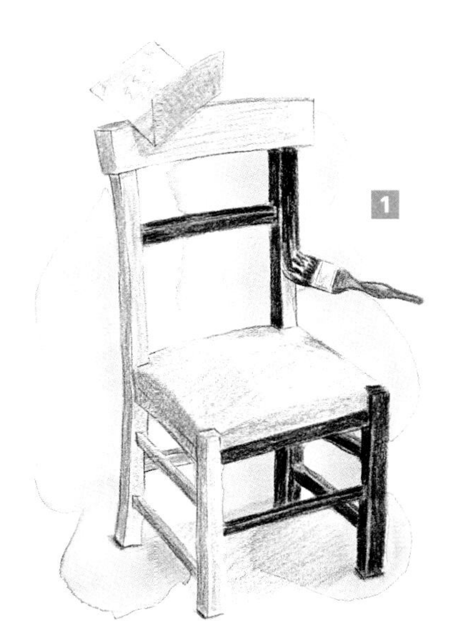

1 Rub the chair down with sandpaper then apply two coats of black emulsion (latex) paint, allowing each to dry.

2 Measure the chair back-rest. Divide the measurements by four and draw a scaled-down plan on a piece of tracing paper to lay over potential designs.

3 Select a coloured background paper and have the designs enlarged onto it via a photocopier to the measurements of the original chair. If there is to be a black border of paintwork around the pasted-up photocopies, the enlargements should be slightly smaller.

4 Cut out all the photocopied pattern elements from the coloured paper. Fold the pattern for the back-rest panel loosely, then make a small crease at one edge to mark its centre. Measure the chair back-rest itself and mark its centre with a pencil dot.

5 Apply wallpaper paste to the back of the photocopy for this panel and to the chair itself, then smooth the photocopy in place, aligning the crease with the pencil dot in the centre. Use a soft cloth to smooth out any air bubbles and wipe away excess paste. Paste and apply the chair and back struts in the same way, allowing a similar proportion of black border around all the pieces.

6 Once the wallpaper paste has dried, apply one or more coats of clear varnish to the whole chair, allowing each coat to dry before brushing on the next.

MEDICINE CABINET

A MEDICINE CABINET IS something that ideally needs to be identified by a bold, graphic, internationally recognized symbol. Everyone, everywhere knows that when you hurt yourself you go to the box with the red cross on the lid, but it doesn't preclude a degree of design fun.

The serious photocopied characters used on the cabinet here, and their detached limbs, are all illustrations used to demonstrate the best bandaging techniques in early twentieth-century nursing textbooks and first-aid manuals. Treatments have changed, and so have the materials used to cover and protect injuries, but these illustrations make an amusing, relevant border to frame our ubiquitous red cross, which is also cut out of paper.

This is really a *découpage* project, demanding tricky cutting out with sharp scissors and pasting up to make a pleasing pattern. In serious *découpage,* the idea is to build up coats of varnish (twenty is not considered excessive), rubbing back each time with sandpaper, until you can feel no bump where the paper edge meets the wood. There is no doubt that this cabinet would make a wonderfully unusual family heirloom if you were to prepare it in this way, but it is perfectly serviceable and attractive with a single coat of varnish to make the surface wipeable.

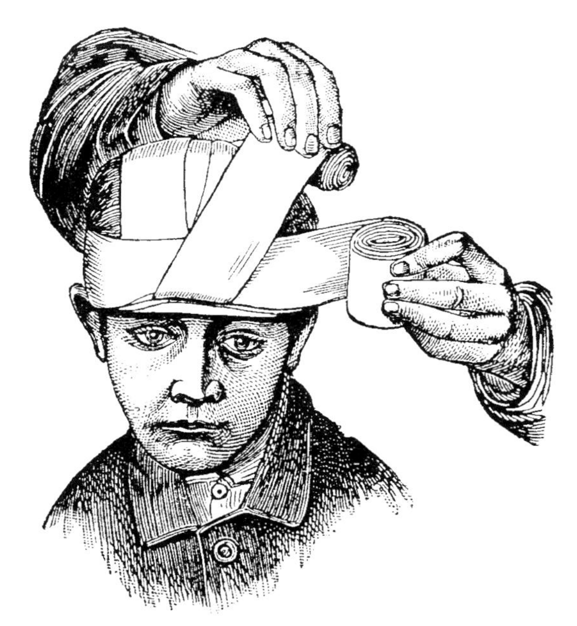

YOU WILL NEED

White emulsion (latex) or satin wood paint
Small wooden cabinet
Sandpaper
Black-and-white photocopies of the wounded and treated (see Sheet 21)
Wallpaper paste
Brushes
Square of thin red paper
Sharp, pointed scissors
Soft cloth or sponge
Clear matt varnish

1 Sandpaper over the whole exterior of the cabinet, then paint it with white emulsion (latex) or satin wood paint.

2 Measure the width of the border area you want to decorate with the photocopied images. If the front of your cabinet is flat, choose a border width that allows enough visual space around the red cross. The idea is to frame the cross, not confuse the signal it gives.

3 Carefully cut out all the photocopied images. Hold the scissors in the same position and move the paper to meet them; this makes it easier to cut into difficult corners without snipping into the images.

4 Mix up the wallpaper paste and coat the outside border area and centre panel of the cabinet. Paste the back of the cut-out images and stick them around the border area, checking that the spacing between them is more or less even. Cut a cross out of the red paper and stick it in the middle of the centre panel.

5 When everything is dry, sponge off any excess paste with a drop of clean water. Then apply as many coats of varnish as your patience will allow, letting each dry before applying the next.

SAMPLER BED-HEAD

IN CENTURIES PAST, little girls were taught essential needlework skills when they were very young and each made a sampler to show off her stitching. Cross-stitched samplers vary in complexity, but they all have great charm. The intense concentration of the needleworker can almost be felt when you look at the colourful little alphabets and other embroidered imagery.

This small bed-head would suit a little girl's room very well, especially one decorated with a touch of nostalgia. The frame around its edge makes the bed-head look like an old picture on the wall, while the enlarged scale of the mesh and stitching brings it right up to date.

The colour copies used here were taken from an actual sampler, rather than a photograph, which accounts for the clarity of the prints. An object loses some of its sharpness when reproduced photographically, then the further processes of photocopying and enlarging will also take their toll. Taking paper copies from a piece of embroidered cloth is clever stuff anyway, so do try and use the real thing if possible.

The photocopies used here are three A3 enlargements butted up and pasted onto a sheet of 9 mm MDF using wallpaper paste. The board was then framed using a picture frame moulding. MDF is a heavy material, and it is important to fix the bed-head securely, either to the existing head-board or to a wall. As beds come in so many different designs, it is not possible to specify the best method of doing this, other than to stress the need for fixings safe enough to withstand the sudden bursts of energy that emanate from happy children.

YOU WILL NEED

Cross-stitch sampler or a good colour photograph of the same
MDF to fit the bed-end (this is a 2 ft 6 in bed)
Wallpaper paste
Brush to apply the paste
Soft cloth or sponge
Moulding for the frame and mitre box (if you want to do this yourself)
Panel pins
2 key-hole mirror plates and screws for fixing the bed-head in place

ABCDEFGHIJKLMNOP
QRSTUVWXYZ

1 Select a sampler, or just a section of one, that has the right proportions to make a bed-head. It needs to be shallow and wide, rather than tall and thin. Have the sampler colour copied and enlarged in sections to the width of your bed. Now measure the height of the enlargement. Have a piece of 9 mm MDF cut to fit this rectangular size.

2 Mix up some wallpaper paste according to the manufacturer's instructions and apply a coat to the MDF and to each of the segments of the enlargements (cut to butt together as on page 44), then stick them down and smooth out any air bubbles with a soft cloth or sponge. Leave to dry.

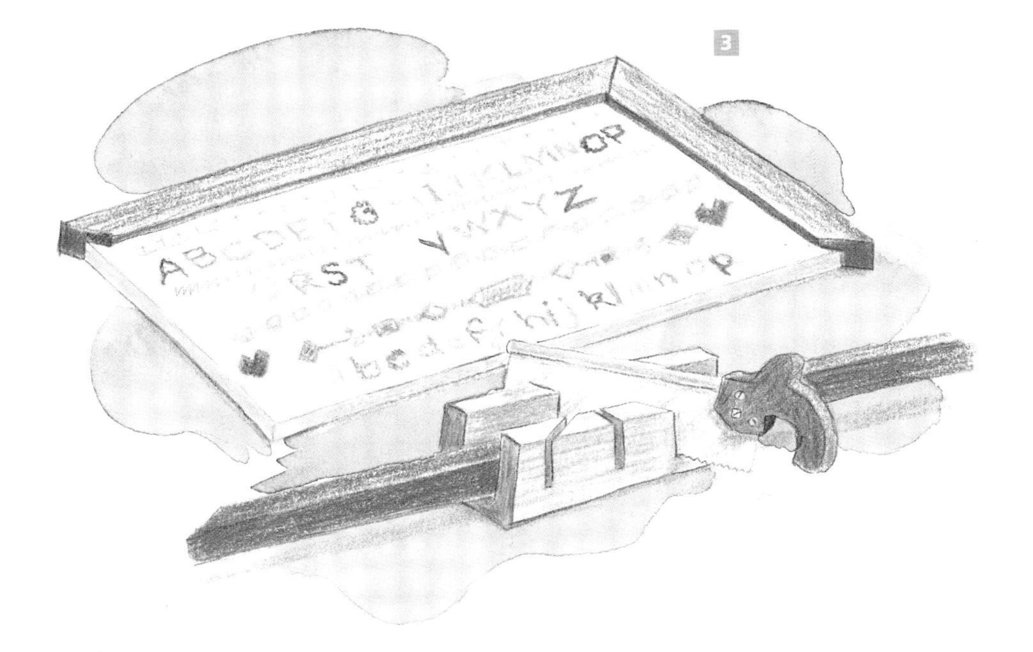

3 If you are equipped and experienced, you can make the frame of the bed-head yourself. If not, take the easy route and ask a picture framer to fit the bed-head with a simple wooden moulding.

4 If your bed has an existing head-board, the new one can simply slot in front of it, with screws at the back to hold it in place. For a divan base, attach the head-board to the wall using two key-hole mirror fixings, and push the bed up against it.

Chapter Five

Decorative Objects

THE PROJECTS IN this chapter are intrinsically different from the others in the book because the photocopied images are the starting-points of the ideas rather than the finishing touches. Each object relies upon the paper copy to make it work, with the possible exception of the tray, which admittedly would be useful for carrying cups even if it had not been made to look like a piece of fancy knitting.

Here you can experiment with the surreal by using photographic prints of textures to give false messages. Feathers, jelly, balls of wool or plates of baked beans are instantly recognized in our minds as having a particular texture, weight and temperature, confirmed by past experience. But when we photograph them and use photocopies of the prints to transform surfaces, we make more than just a pattern: we also set up contradictory messages about the nature of the image and its appropriateness to the functional object. These messages are startling and need to be processed in our minds. The eventual answer comes back that this must either be humour or art; and most of us would be pleased to be responsible for both. Whether you choose to pursue the

visual joke or play games with people's senses, creative thinking is required for these objects, uplifting in itself.

None of the projects in this chapter is difficult to make or do, and their novelty value is based upon the images that you choose and the way the objects are displayed. This is personal expression. Table-mats, for instance, exist in all our homes, but there is undeniable pleasure to be gained by making a set of these very everyday objects yourself. Each time you lay the table there is a sense of satisfaction that comes from having created them rather than just gone shopping. The mats we make here are laminated (see page 106), but you could also paste copies onto plywood or MDF and varnish them, perhaps cutting them into irregular shapes, as we did with the fridge magnets. Similarly, the fridge magnets could be cut out as simpler shapes, leaving the image to create the effect.

The idea of making a dolls' house by sticking an enlargement of a house onto a wooden box uses the photocopier to create illusion. Why not use the same idea in other parts of the home, your kitchen cupboards, for instance? Melamine can be rubbed down with sandpaper to provide a rough surface for the paste to `key' into. Any flat surface has the potential to masquerade as something else – bumpy, spiky or soft and quilted – it's up to you.

All of the ideas in the projects here are interchangeable. They will work well as we have presented them, but you will have additional fun if you take the information you need and use it as a springboard for your own ideas.

PICTURE FRAMES

Although frames started life in a supporting role, as a means to display and protect the art within, they have also been elevated to artworks in their own right. People buy frames for their own sake now, then look for something to put inside later. Stylish frames can be expensive, but you can customize bargain buys using photocopied material and change them into classy accessories. You can buy bargain frames in end-of-season sales, or old ones from boot sales and junk shops.

We have given these three plain, flat, wooden frames very different treatments. One has a very bold, graphic look, which comes from using a sheet of letters from a book of typefaces. Letter forms are designed with awareness of the effect of positive and negative space – so there is good balance between the black and white areas, even when you enlarge them and use them for their own sake rather than for spelling out a word. The mitred corners help to enforce the purely decorative use of the letter forms.

The second frame has been covered with a border design which you will find on Sheet 2. This pattern has been taken from a Greek frieze and will add a touch of historical elegance to any room. The black-and-white copier paper has been coated with a 'light oak' tinted varnish, which has turned it a lavender colour and added a slight sheen. To retain the original black, grey and white, just use a clear varnish instead; you can also vary the sheen by choosing from a matt, satin or gloss varnish. The pattern strips have been cut so that the sides are covered first with the top and bottom strips overlapping them slightly, with no mitring needed.

The coloured frame is the simplest one to do. This is a photograph of an old patchwork quilt called 'tumbling blocks' which has a wonderful three-dimensional quality. The wooden frame was first painted to match one of the quilt colours and the print has been cut slightly smaller than the frame so that the background colour shows around the edges. A coat of clear varnish adds to the jewel-like brilliance and helps to bond the paper to the frame so that no corners lift.

Whether you take you own photographs and have them photocopied or just find printed images you like in books, magazines or newspapers, you can change a pretty boring frame into something special in the space of half an hour.

YOU WILL NEED

Flat wooden frames – curves and mouldings are tricky, so begin with an easy shape
Wallpaper paste
Old brush for the pasting
New, or very clean, flat ended brush to apply the varnish
Your choice of varnish – matt, satin or gloss, tinted or clear
Photocopied images
Ruler
Pencil
Scalpel

THE MITRED FRAME

1 Photocopy an alphabet from a letter-form source book. Enlarge to get a bold black-and-white pattern.

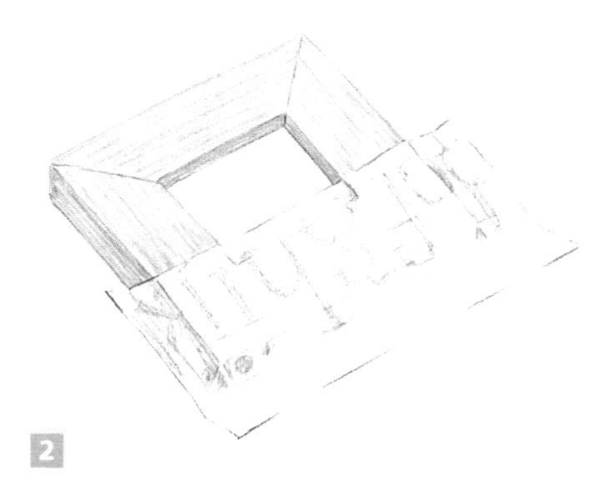

2 Measure the frame, then cut the strips, allowing for overlaps. You need to measure across the top and down both the outside and inside edge, then add on an inch or so (a couple of centimetres) for the overlaps – each frame is different, so no exact measurements can be given.

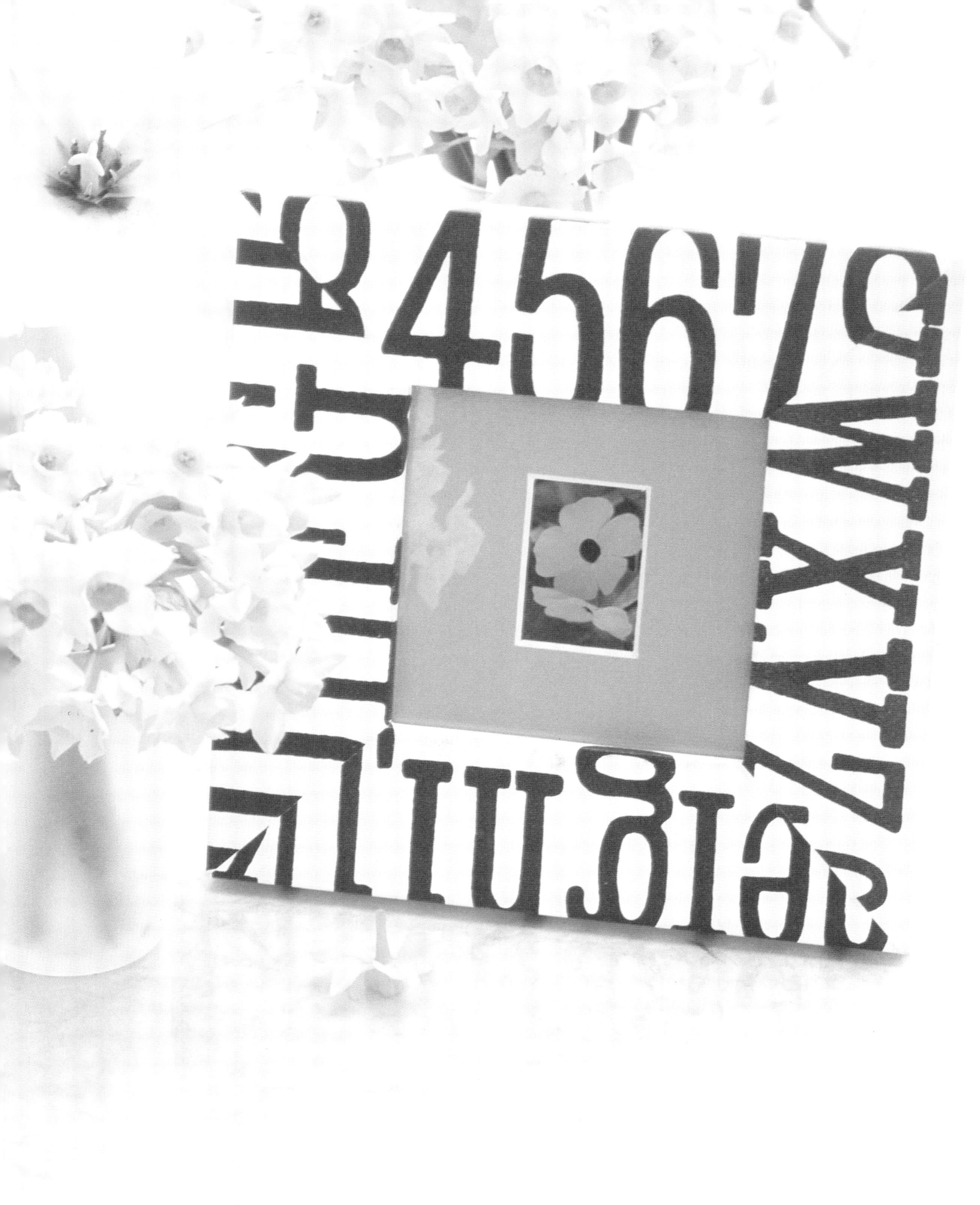

3 Use the frame itself to make your measurements. Place the strips on it and mark the inside corners and outside edges by pressing your fingernail into the paper, but not through it. The mitred corners should be cut so that they overlap the corners slightly on the first two opposite strips; then you can cut the final two to fit exactly into the corners, for a very neat finish.

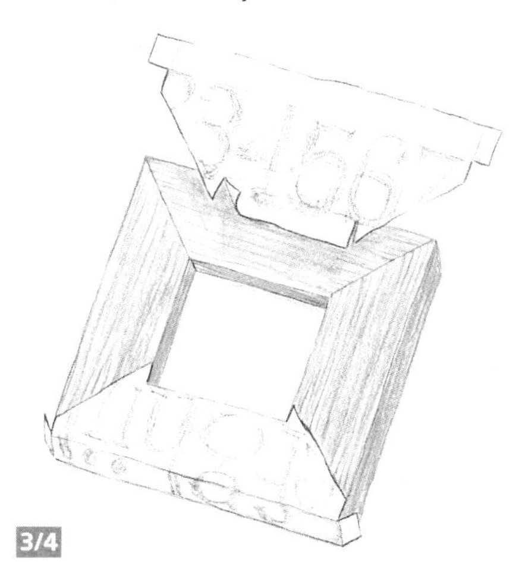

3/4

4 Apply an even coating of wallpaper paste, then smooth the top down, tucking the inside flap in, and smoothing flat. Smooth the side down and press the two ends around the corners.

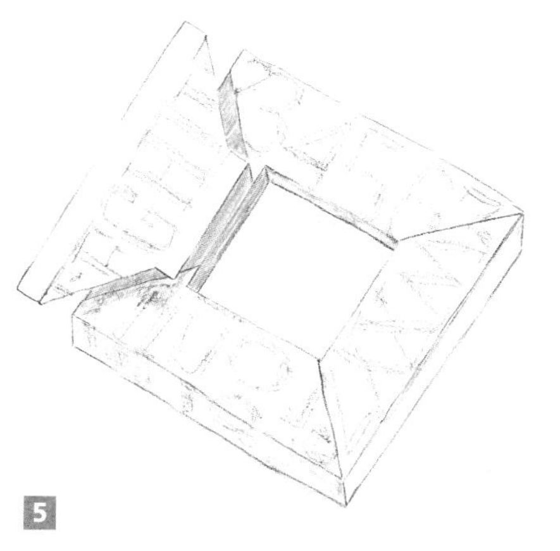

5

5 Paste and apply the second pair of strips in the same way, but line up the mitred edge with the inner and outer corner this time. These strips will cover the first overlapping corner edges, which is easier than cutting two edges to meet perfectly. Varnish.

THE CLASSICAL FRAME

1 Photocopy the patterns that appear on Sheet 2. Cut these into strips wide enough to cover and overlap the top, inside and outside edges.

2

2 Place the frame face down on the top and bottom strips, marking their position in pencil, as in the illustration. Use a scalpel to cut two short slits on one edge to make the inside edge fold; and remove two triangular notches that will allow the sides to fold around neatly.

3 Use the frame to measure the size of the two side pieces. They are simple rectangles that fold inside, over the top and down the sides.

3

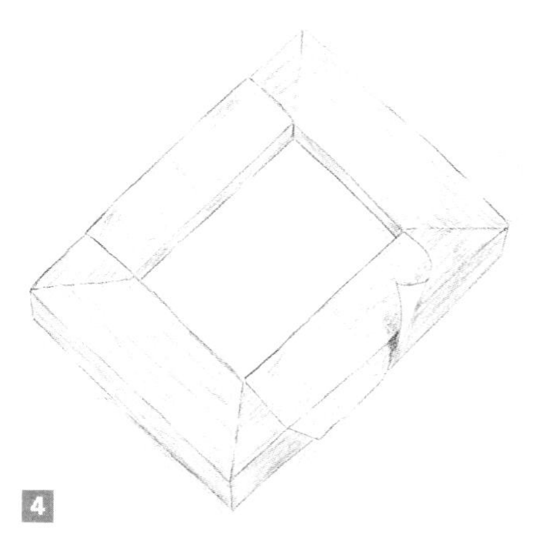

4

4 Coat the two side pieces and the area to be covered with with wallpaper paste and smooth the side papers onto the frame.

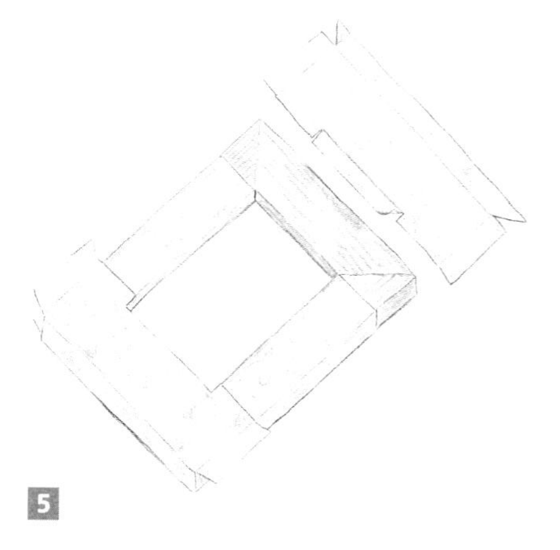

5

5 Do the same with the top and bottom strips, tucking the triangular ends around the corners, then covering them as you fold and smooth down the side pieces. Varnish – we have used a 'light oak' tint.

THE PATCHWORK FRAME

1 Make a colour copy to use on the frame. This could be of a quilt, embroidery, an old map – anything that catches your eye.

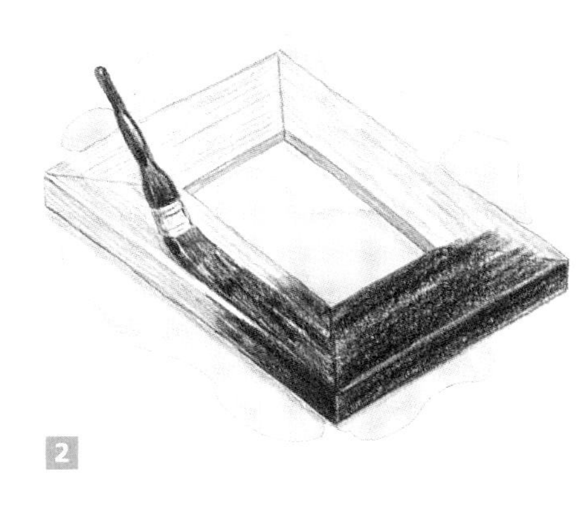

2

2 Paint the frame in one of the predominant colours of your print. Allow to dry.

3

3 Place the frame face down on the copy and draw around the shape.

4

4 Cut out the shape, then trim 1 or 2 inches (2.5–5 cm) from both the inside and outside edges.This will give you the small border of colour around the edge; adjust the width according to the size of your frame.

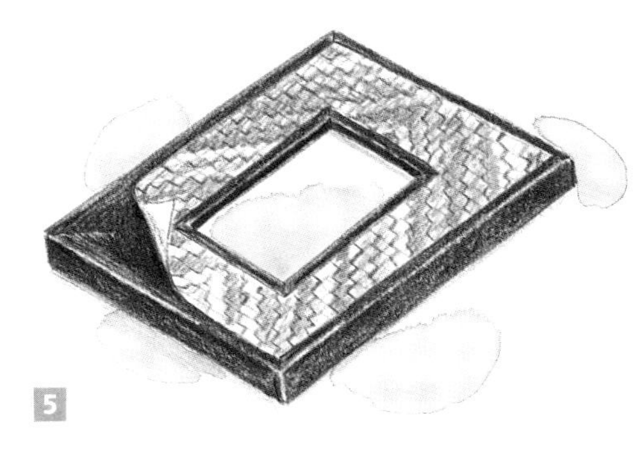

5

5 Apply an even coating of paste to both the frame and the paper, then stick it in position and wipe off any excess that oozes out as you flatten the print.

6 When the paste is dry, you can apply a coat of varnish to seal and protect the frame.

'KNITTED' TRAY

TRAYS COME IN ALL shapes and sizes, but the qualities that we demand of a tray are chiefly that it is both rigid and flat. A knitted woollen tray would be pretty useless, which is probably why the appearance of our tray here is so appealing, almost surreal.

You don't even have to take a photograph or find a picture of knitting to copy: you can take the real thing along to the colour photocopier and make a print straight from it. This one is taken from the back section of a child's Fair Isle cardigan, but you could use a plainer knitted pattern in a single colour or even some fabric, like denim, corduroy or hessian, that has a very visible weave and texture. The basic idea is the same: to play a visual joke but to play it with style, so that your quirky little tray remains in use long after its novelty value has worn off.

YOU WILL NEED

Tray to cover – choose a simple shape without nooks and crannies
Wallpaper paste and brush to apply it
Ruler
Scalpel
Clear varnish and brush to apply it
A3 photocopy (2 if your tray is a large one) of a piece of knitting or fabric
Felt to cover the base (optional) and PVA or clear glue to stick it down

The instructions here are for covering a tray with cut-out handles. Avoid trays with curved mouldings – at least until you've mastered a simpler shape.

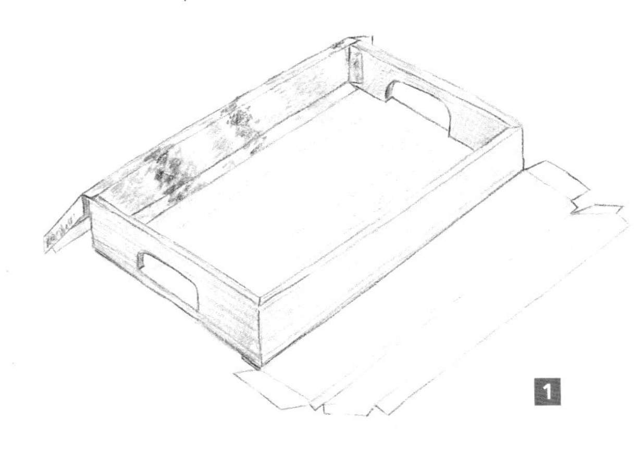

1 Use the tray to measure the width and length of the side pieces. Place it on the photocopy and allow half an inch (1.25 cm) for folding under, then add the height of the side, the flat top, the inside flat edge and another half inch (1.25 cm) onto the inside base. Extend the width by about an inch (2.5 cm) each side to make flaps to fold around the ends. Use your fingernails to mark the paper with slight indentations to use as cutting guidelines.

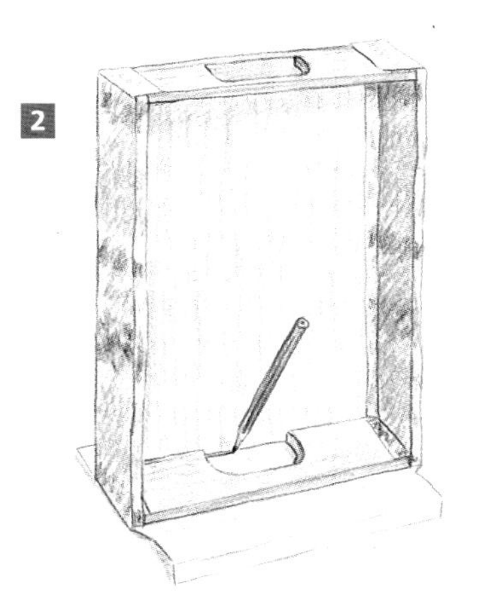

2 Measure the end pieces in the same way. Once you have cut out the two pieces, upend the tray and mark the position of the handle. Don't cut the shape out. Use a scalpel to cut a line along the middle of the shape, then make cuts radiating out into the curves to meet the edge of the pattern shape. When you paste them down, these will fold inside the cut-out shape.

3 Apply paste to the side pieces and the sides of the tray, then smooth them into position. Wipe off excess glue.

4 Apply paste to the end pieces and the ends themselves. Starting on the outside, smooth the pieces into place, so that the cut is aligned with the hole. Use your finger to smooth the flaps of paper into the handle. Smooth the paper over the top and down the inside, then onto the base. As pasted paper is notoriously difficult to cut when wet, leave the cutting of the inside opening until later.

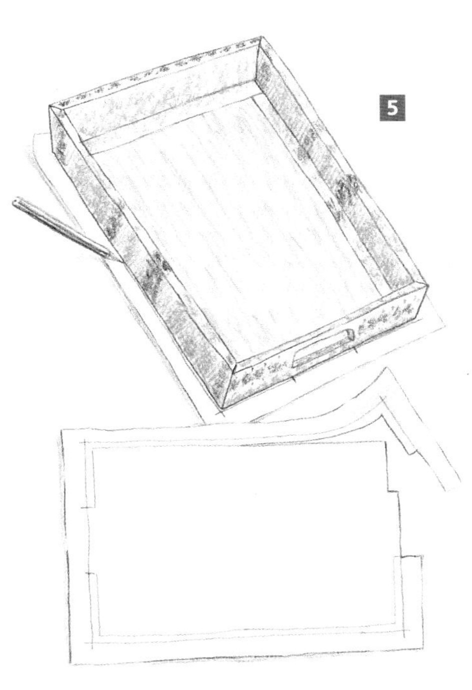

5 Place the tray on top of the print and draw around it. Subtract the depth of the sides to find the dimensions of the inside base (or make a paper pattern by folding paper to fit, if you find it easier).

6 Paste the base of the tray and its covering paper, then smooth it into position to cover the overlaps from the side and end pieces.

7

7 When the tray has dried, hold a piece of card butted up against the inside tray end to support the paper and prevent it from tearing. Use a scalpel to cut the slits for the handle, in the same way – but not in exactly the same positions – as before. The idea is to cover any gaps with these flaps.

8 Apply paste to both surfaces and use your finger to press and smooth the paper into the handle from the inside of the tray. The paste and paper will take up the shape of the hole as they would in *papier-mâché* work. Leave to dry for a couple of days and then apply several coats of clear varnish.

9 The base of the tray can be covered with felt, cut to size then glued in place with clear glue or PVA.

WRISTWATCH WALL CLOCK

IF YOU HAVE never made a clock before, you will be forgiven for thinking of it as an area where specialized knowledge is required. Nothing could be further from the truth. Battery-operated clock movements are not only cheap, but also reliable, and a single drilled hole is all that's needed to fit the parts together with the works behind and the hands on the front.

This wristwatch was copied from a 1940s department store catalogue and then given the Pop Art treatment by enlarging it way beyond life-size to make an unusual wall clock. One of the joys of photocopying is being able to play around with scale in this way, and it works particularly well when fine engraved lines are enlarged to reveal their irregularities. The results are bold and graphic.

The first stage of the project involved making an A3 black-and-white enlargement. This became the master copy and needed some retouching before moving on to the next stage. A black-and-white image can be printed on a colour photocopier with a single chosen colour replacing the black. The watch was printed in brown on a light cream paper. The print colour range is large and a copy shop will have a selection of different coloured papers to choose from – or you can take your own along, trimmed to A3-size so that it fits the machine.

Four copies of different sections were needed to make up the wristwatch. They were pasted onto a sheet of MDF, and once dry, the shape was cut out with a jig-saw. Coloured inks provide brilliant transparent colour that is perfect for this type of print – turquoise, yellow and orange were used, but once again, there is a large colour range to choose from. When buying the clock movement, you will need to know the depth of the MDF, then it is just a matter of drilling a hole and snapping the two parts into position.

YOU WILL NEED

Print of a wristwatch
Scalpel
Cutting mat or card
Black marker pen
Sheet of 9mm thick MDF - ask for an off-cut
Masking tape
Wallpaper paste
Brush
Coloured inks
Small artist's brush
Clear varnish
Battery-powered clock movement
Drill
Jig-saw
G-clamps
Acrylic paint for the cut edge of the MDF

1 Have an A3 black-and-white enlargement made of the wristwatch.

2

2 Use a scalpel to cut away the clock hands and the marker pen to touch up any big gaps and to strengthen weak lines. Take care not to smarten the line up too much, or it will lose all its character.

3 The next stage is done on the colour photocopier, using off-white paper and replacing the original black with brown. Ask the copier operator to enlarge the A3 print to the finished size you want, but explain that the resulting sections will be pasted up to make one large image. The copies you get should include an overlap, so that adjoining areas appear at the edge of both sheets.

4 Arrange the four copies to make one perfect image by placing them one on top of the other at the overlaps and securing them in place with low-tack masking tape. Cut through with a scalpel so that the edges butt up perfectly. Apply wallpaper paste to the MDF and the back of the prints as you stick them down in position. Smooth out any air bubbles and leave to dry.

4

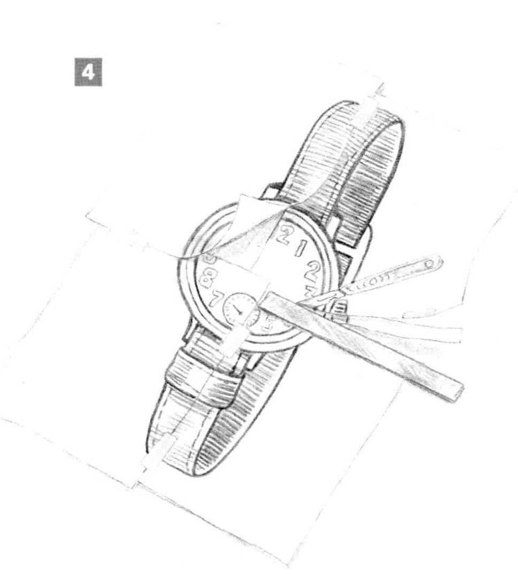

5 Clamp the MDF to a workbench or table-top and use a power jig-saw or a coping saw to cut around the shape. Paint the cut edge of the MDF with brown acrylic paint.

5

6 Colour the print using inks and an artist's brush and apply clear varnish when dry.

7 Drill a hole through the middle to take the movement. Check the size: ours was made using a 10 mm bit, but designs do vary in size. Fit the two parts together, set the time and hang your new Pop Art clock on the wall.

6

FRIDGE MAGNETS

FIFTIES DESIGN WAS a celebration of the future and of freedom from the austere forties – out went drab and in its place came brilliant colour, streamlining and an explosion of luxury consumer goods. Magazines were packed with adverts that urged everyone to change their lives and join the modern world by installing the very latest electrical, labour-saving devices and driving around in long, sleek, chrome-lined cars.

Something that these design pundits would never have dreamed up is fridge magnets – the very idea of spoiling the smooth lines of those fabulous refrigerators would have seemed like sacrilege. These days, most of us find them indispensable. The fridge has become a noticeboard – a place to stick up reminders, invitations, photographs, cartoons and messages. We know we'll be going to the fridge frequently, so they won't be missed.

We plundered some fifties magazines to come up with the images used on these fridge magnets – they have that contemporary cool look you sometimes get when you take high style from one era and move it on a few decades.

YOU WILL NEED

Colour photocopies
An off-cut of 2.2 mm MDF
Coping saw
V-board – make one of these by literally cutting a 'V'-shape out of a piece of MDF, as described in Step 3
2 screws to hold it in place, or a G-clamp
Wallpaper paste and brush to apply it
Clear glossy varnish and brush to apply it
Fine sandpaper
Magnets – buy small, flat, circular magnets from hobby and craft shops
All-purpose clear glue

1 Search out old magazines, cookery and home improvement books from the early fifties (junk shops, jumble sales and flea markets are good places to look). Select images and have them colour copied and enlarged or reduced to the right scale. When deciding on the size, bear in mind that the magnets need to be powerful enough to hold the MDF and the papers you are going to be displaying, whilst withstanding the jarring of an opening and closing door – so keep the images on the small side and buy strong magnets.

2 Separate the shapes and use wallpaper paste to stick them onto the MDF. Leave until bone dry.

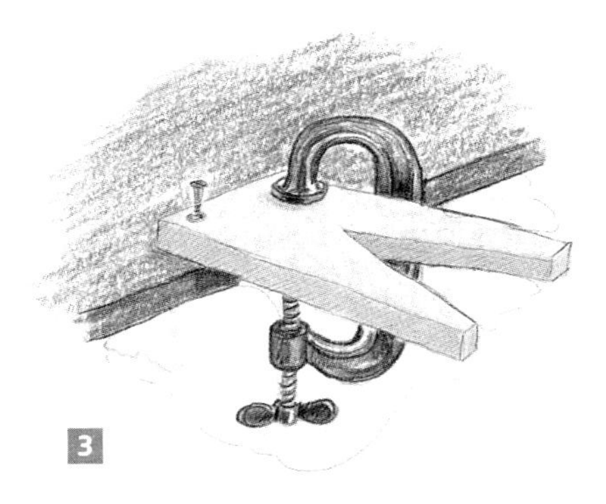

3 To make a V-board, you need a piece of MDF or wood measuring 8 by 4 inches (20 x 10 cm). Cut a 5 inch (12.5 cm) 'V' out of one end. Ideally, this should be screwed onto the edge of the workbench so that it overlaps the edge by 6 inches (15 cm). If you use a clamp to hold it in place, it will have to be moved around to allow ease of movement for the saw.

4 Cut out the shapes roughly at first, so that you have smaller shapes to work with.

5 Place a shape on top of the V-board, holding it down with one hand. Use the saw vertically, moving the shape around on the board so that the blade of the saw can remain in more or less the same position (as it would if you were using the equivalent power tool, the band saw). When sawing into a sharp corner or bend, don't cut in then try to turn the blade and cut outwards: instead, approach it from two different angles so that the two cuts meet each other, then remove the waste.

6 Lightly sandpaper the edges of the cut-out shapes to smooth them, but take care not to lift the paper when you do this.

7 Apply two or three coats of varnish, allowing adequate drying time between coats.

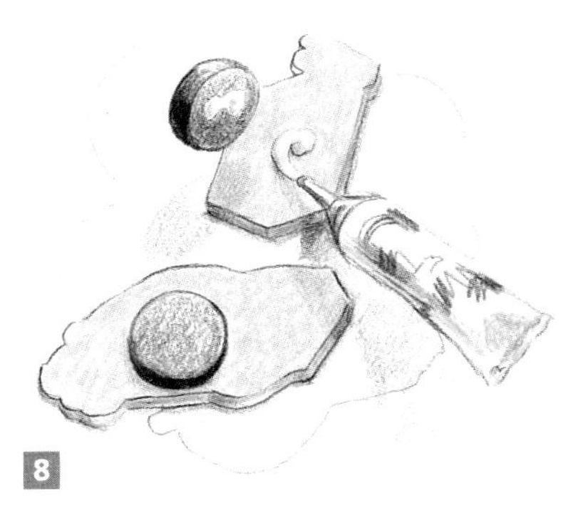

8 Use the clear glue to stick the magnets to the centre backs of the shapes.

TABLE-MATS

THE RANGE OF table-mats in the shops today has certainly come a long way since the days when the choice was between florals, historical scenes or Old Master paintings, yet it is still difficult to find anything other than the current trend in colours and designs. But once you realize how easy it is to make your own, you will never, never buy another.

The secret lies in the laminating service offered by copy shops – it means you can choose absolutely any image you like, photocopy it and have it sealed between two sheets of clear, plastic film to make the most individual table-mats imaginable.

Our table-mats have been made by colour copying photographs featured in a book of old American quilts. Each one is different, but the textile theme binds them together as a set. The fact that you are doing something for your own use, not publication or re-sale, means that you can probably risk taking copies from books, magazines, newspapers or advertisements without incurring any wrath over copyright. It would be a strange, brave new world if our homes were inspected for breach of copyright laws!

Once you have selected your image or images, take them to the copy shop and have A3 colour copies made. Trim off the white borders and ask for them to be laminated in the heavier material known as '10 thou'. This will give a strong, wipeable coating that will protect the table-top and be able to withstand the heat of hot-plates and dishes – although it is not recommended for pans and dishes straight from the oven or hot plate.

The cost is relatively low, so you may find yourself making sets for special occasions. Why not serve your Indian meals on temple carvings or vibrant sari material, and personalize your Christmas place-settings with a photo of your dog in a party hat? Fussy toddlers may be persuaded to eat up if dinner is served on the right mat – a collaged photocopy of themselves with their favourite cartoon character or TV presenter, perhaps. This project is so easy that it doesn't need steps to show you how to do it – you make the design decisions and the copy shop does the rest.

DOLLS' HOUSE

ALL CHILDREN LOVE dolls' houses, not just girls. Little boys may squirm a bit at the idea of actually owning one, but if a house just happens to be there and they come across it, they will spend many happy hours rearranging the furniture.

Buying and furnishing a dolls' house can be so expensive that you may feel reluctant to let a small child anywhere near it! This neat idea will solve the problem – it is a sturdy box, divided into four 'rooms' around the back that are virtually impossible to wreck. The simplicity of the house will encourage creative play, because all the decision-making is left to the child.

The façade is a flat sheet of MDF covered with a photocopied line drawing of a house. It is a bit like one of those fake towns they make when filming Westerns. The simple shape of the rooftop and chimney can be cut out if you have a jig-saw, but its complexity will depend upon the design of the house. You could even use a photocopied enlargement of a photograph of your own house on the front.

Paint the house, window frames and front door, adding as much detail as you can to bring it to life: roses around the door, window boxes, curtains in the windows and cats on the ledges. A coat or two of varnish will seal and protect the surface, then you should brace yourself and hand the decorating over to a small child! The instructions here are only a guide, as scale and proportion may differ in the case of your own house.

YOU WILL NEED

- One inch (19 mm) MDF cut to the following sizes:
- 2 pieces measuring 10 x 16 in (25 x 40 cm) for the sides
- 2 pieces measuring 10 x 25 in (25 x 63 cm) for the top and base
- 2 pieces measuring 10 x 7½ in (25 x 19 cm) for room dividers
- Quarter inch (5 mm) MDF measuring 27 ½ x 16 in (68.5 x 50 cm) for the façade
- Jig-saw (power tool) or a coping saw
- Drill
- 12 screws
- Wood glue
- Wallpaper paste and brush
- Sponge or soft cloth
- Sandpaper
- Photocopied black-and-white enlargement of the front of a house
- Low-tack adhesive tape
- Glue gun
- Hand saw
- 2 G-clamps
- Sturdy workbench or table top
- Small hammer
- Panel pins

TO MAKE THE FAÇADE

1 Choose a picture of a house for the façade – it can be a line drawing or a photograph. This comes from an old builders' plan source book.

2 Give the measurements of the façade to the operator and ask for enlargements to that size. This took nine A4 copies. When enlargements are made in sections like this, theoverlaps mean that you can get perfect alignments.

4

4 Mix up wallpaper paste. Paste the board and the photocopies, then stick them down, starting at the bottom edge and smoothing the paper as you go to remove excess paste and air bubbles.

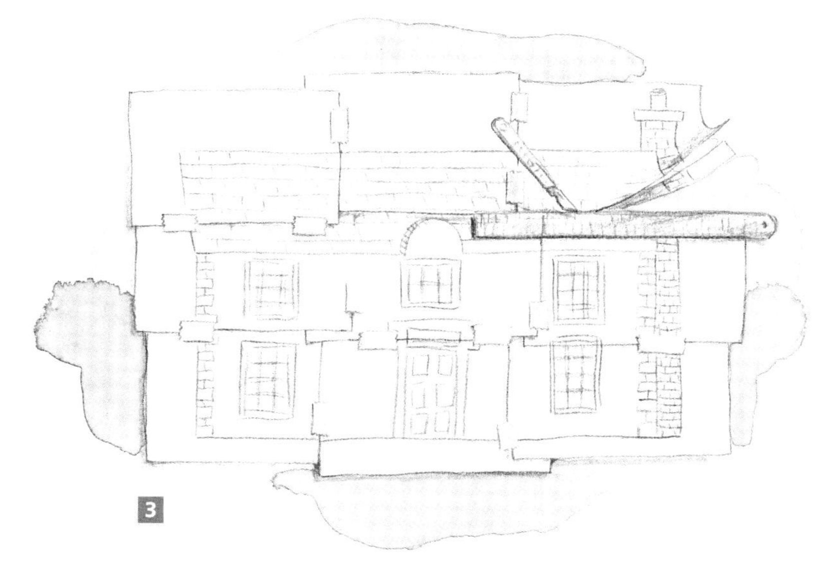
3

3 Lay out all the copies in perfect alignment. Join them together with small pieces of low-tack tape (ordinary tape will rip the paper when it is removed). Now place a steel ruler along the middle of each tiled section (the overlap) and cut through both pieces. This way the edges will meet perfectly.

5 When the paste has dried, the MDF is clamped to a work surface to cut out the chimney and the roof slope, using a jig-saw. You have to keep the blade vertical and the foot plate in touch with the surface: otherwise it can vibrate and cut in a less-than-perfect line. Sand the edges lightly.

5

TO CONSTRUCT THE BOX

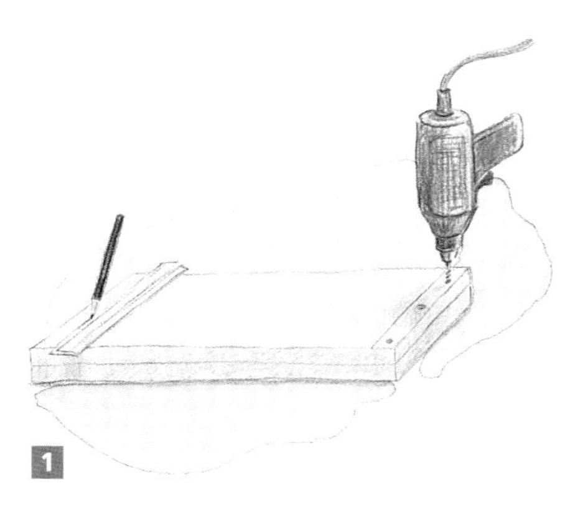
1

1 The box is joined together with screws. Draw a pencil line an inch (2.5 cm) inside each end of the longer pieces. Drill three holes, spaced along the middle of this narrow section. They will take the screws that hold the sides in place.

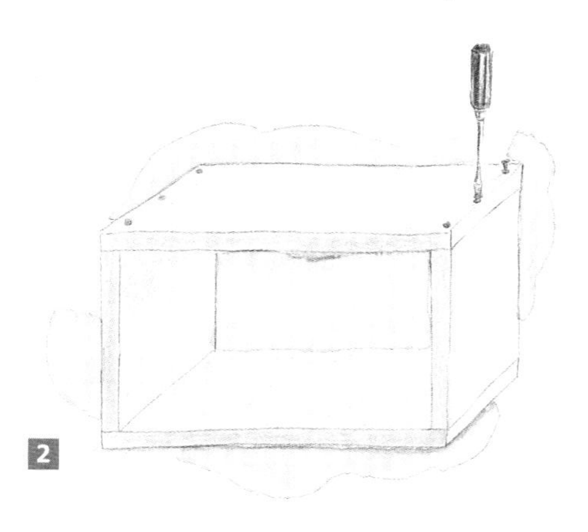
2

2 Screw the four pieces together to make the box.

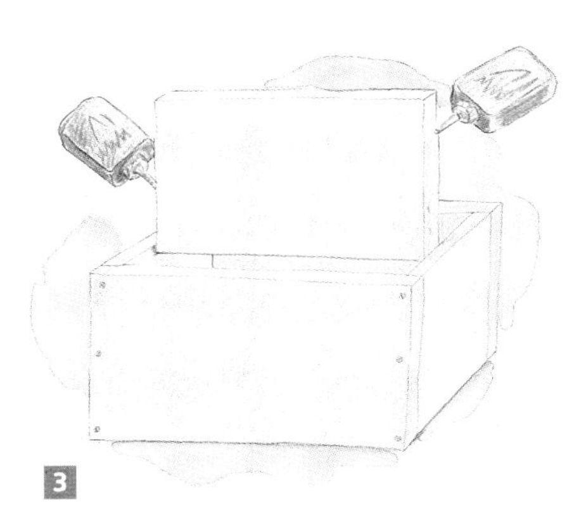
3

3 Apply wood glue·to the ends of the horizontal divider and slot it into place.

4/5

4 Apply wood glue to each end of both the vertical dividers and slot them in place. These should fit snugly and not need supporting in any way while the glue bonds. If by chance there is some room for movement, you cold run a strip of beading along the base, which would look like a skirting board.

5 Tap a few panel pins into the dividers from above, below and at the sides.

TO PAINT THE DOLLS' HOUSE

YOU WILL NEED

Windsor & Newton's Brilliant Watercolour Paints (in bottles with droppers)
Acrylic gloss varnish
Coloured inks and a brush
Artist's broad square-tipped brush
Smaller brush for details
Pale grey emulsion (latex) paint

1 Mix pink for the walls by adding a dash of brown into cherry red, then mixing the colour into some gloss varnish. Test the colour, which should be quite pale. Paint the walls.

2 Mix 1 part blue (process cyan) into 5 parts gloss varnish and paint the window frames.

3 The door panels are burnt sienna and the frame is a mixture of cyclamen and blue hyacinth, which makes a deep burgundy red. As before, these are mixed into gloss varnish.

4 The roof is painted pale grey using a mixture of Prussian blue, burnt umber and white mixed into gloss varnish. The rest of the box is painted with pale grey emulsion (latex) paint. Emphasis and added detail can be given to the outside of the house by touching up with coloured inks and a fine brush. Here, we have picked out some of the brickwork.

5 The front façade is simply fixed to the front of the box with panel pins, and the rooms are accessed from the back.

AUTHORS' ACKNOWLEDGEMENTS

We would like to say a very big 'thank you' to Sam, Janet and Tony at Prontaprint in Hastings, Sussex, for their invaluable help with this book. They were completely impossible to faze, even with our most bizarre photocopying requests. Special thanks to Tony for explaining the fundamentals, as well as the technicalities. Thanks, also, to everyone at our 'local' photocopy shop, Fast Print and Design, in St-Leonards-on-Sea; they always rose to the occasion.

We would probably never have written the book without the enthusiasm of Piers Burnett at Aurum Press, who kept the idea alive; and would not have done such a good job without the support and encouragement of our editor, Judy Spours.

A big 'thank you' to Stephanie for the styling of our projects and to Michelle for her lovely pictures. And to Vikki, for all that tea!

The authors and publishers would also like to thank DYLON International Limited for supplying Colourfun Image Maker for the fabric projects in the book. DYLON operate a consumer telephone advice line and will send out leaftlets on request: 0181-663 4296.

PHOTOGRAPHIC ACKNOWLEDGEMENTS

All original colour photography is by Michelle Garrett.

The black-and-white images reproduced on the following thirty-two pages, and others used throughout the book, have been selected from a wide variety of sources. We are indebted to The Mary Evans Picture Library, London, for permission to reproduce sixty-two images; we are also grateful for permission to reproduce images from The Harrods Archive, London, and from The Garden Archive, London.

Thanks are also due to The Kobal Collection, London, for permission to reproduce the film stills used on the video storage boxes on page 32; and to The Advertising Archive, London, for permission to reproduce the images used on the fridge magnets on page 102.

DETACHABLE SHEETS OF PHOTOCOPY IMAGES

SHEET 1 Borders and Ornaments

SHEET 2 Borders and Ornaments

SHEET 3 Patterns

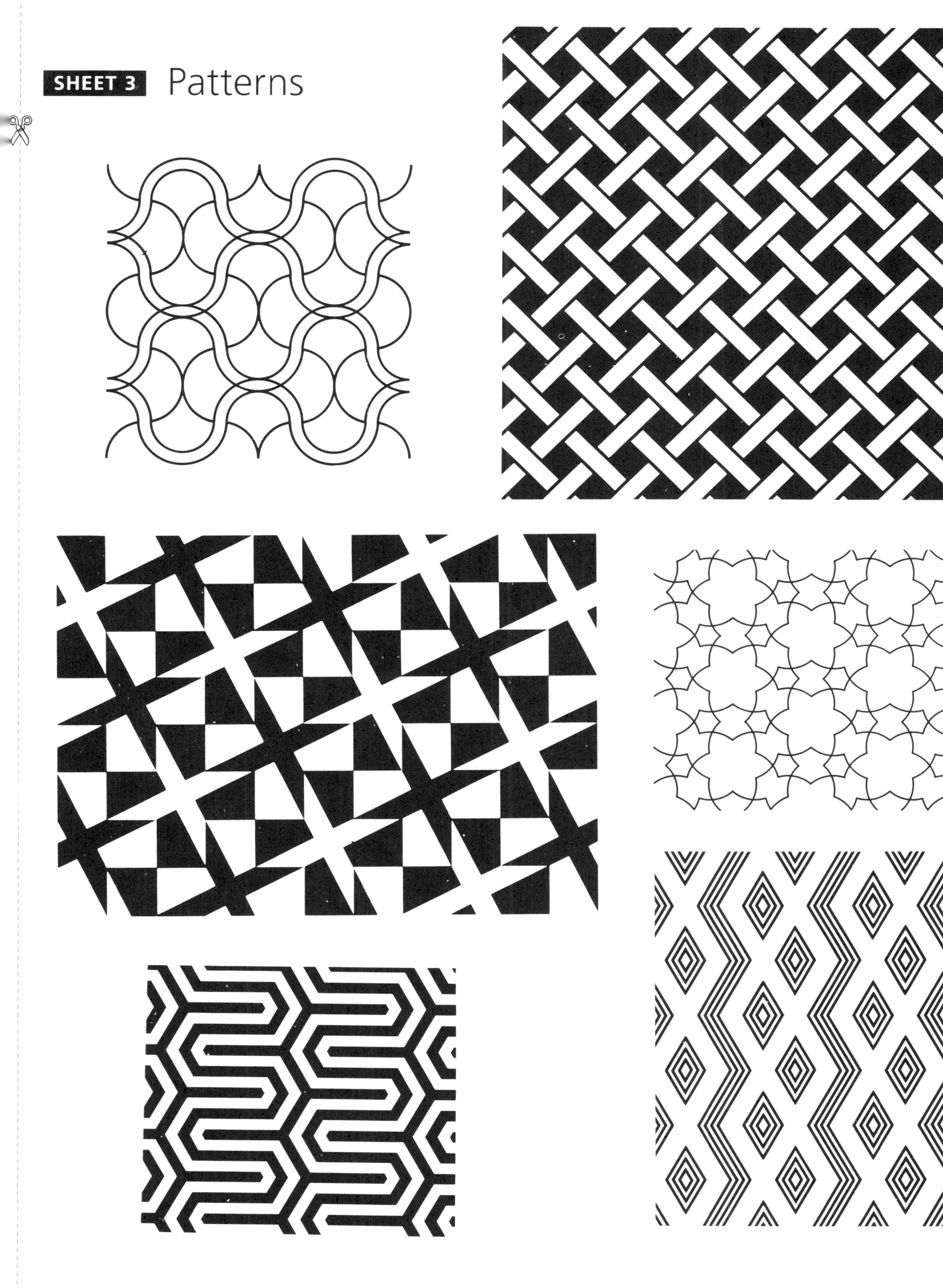

SHEET 4 Frames

SHEET 5 Frames and Banners

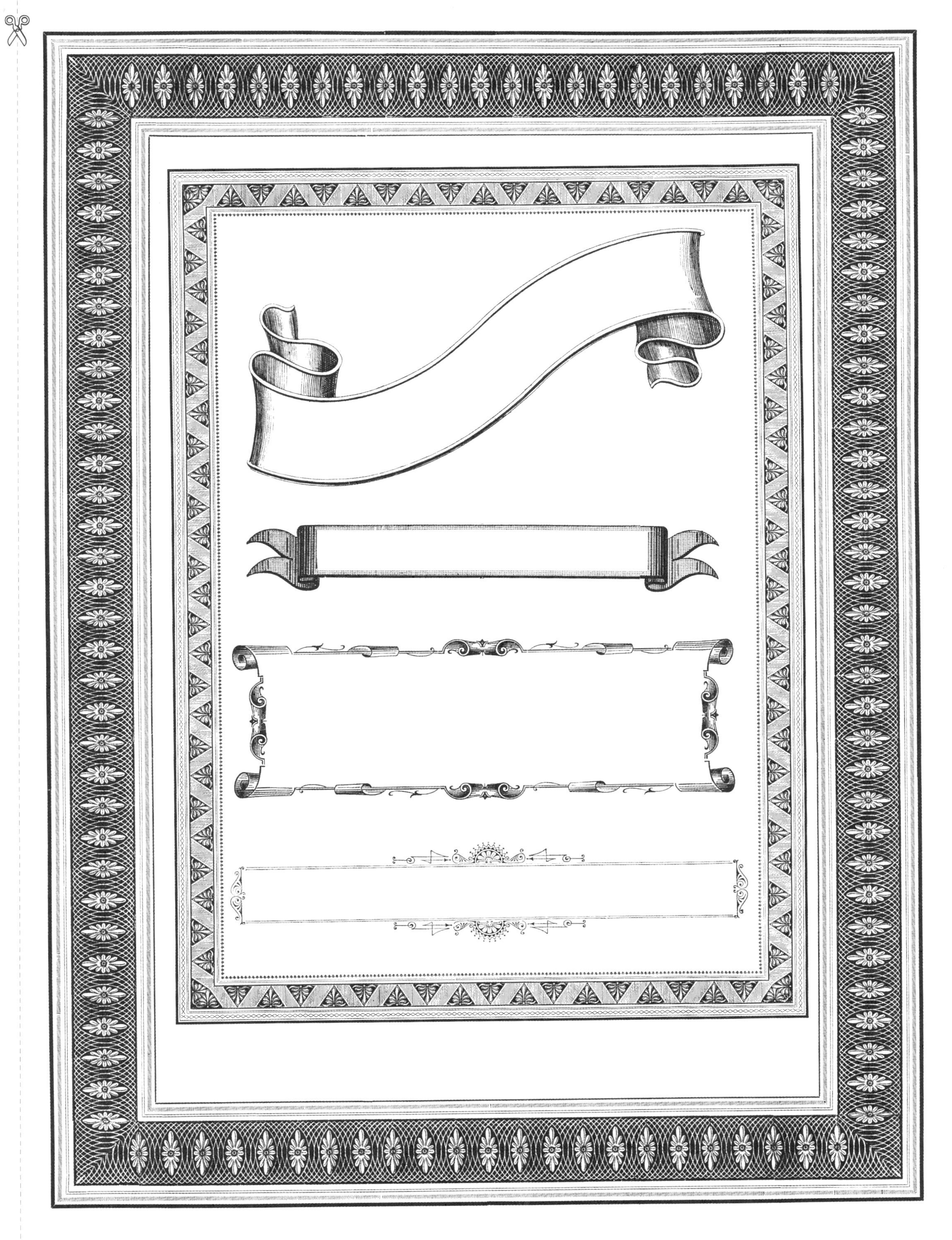

SHEET 6 Frames and Mirrors

SHEET 7 Borders and Mirrors

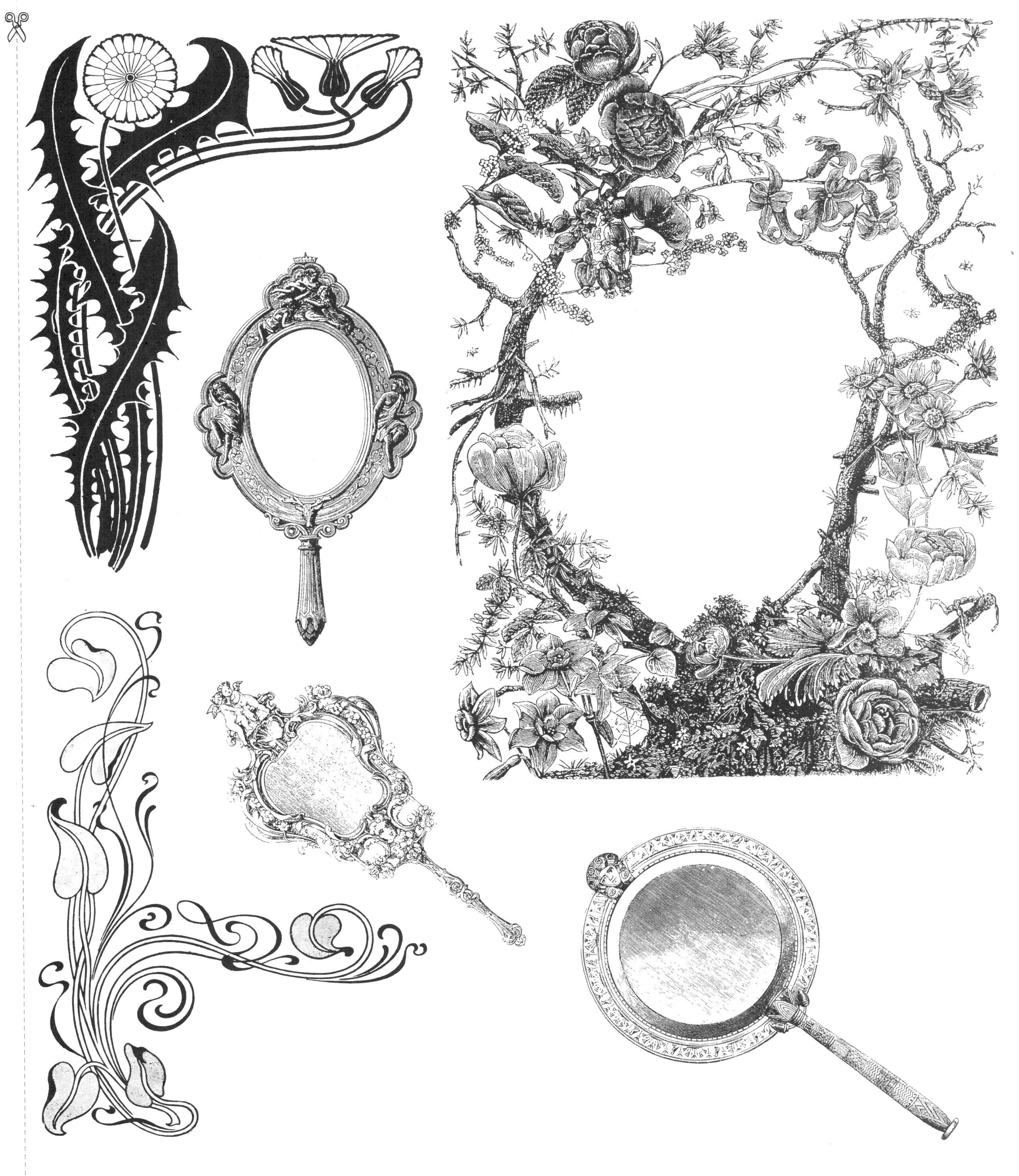

SHEET 8 Fruit and Vegetables

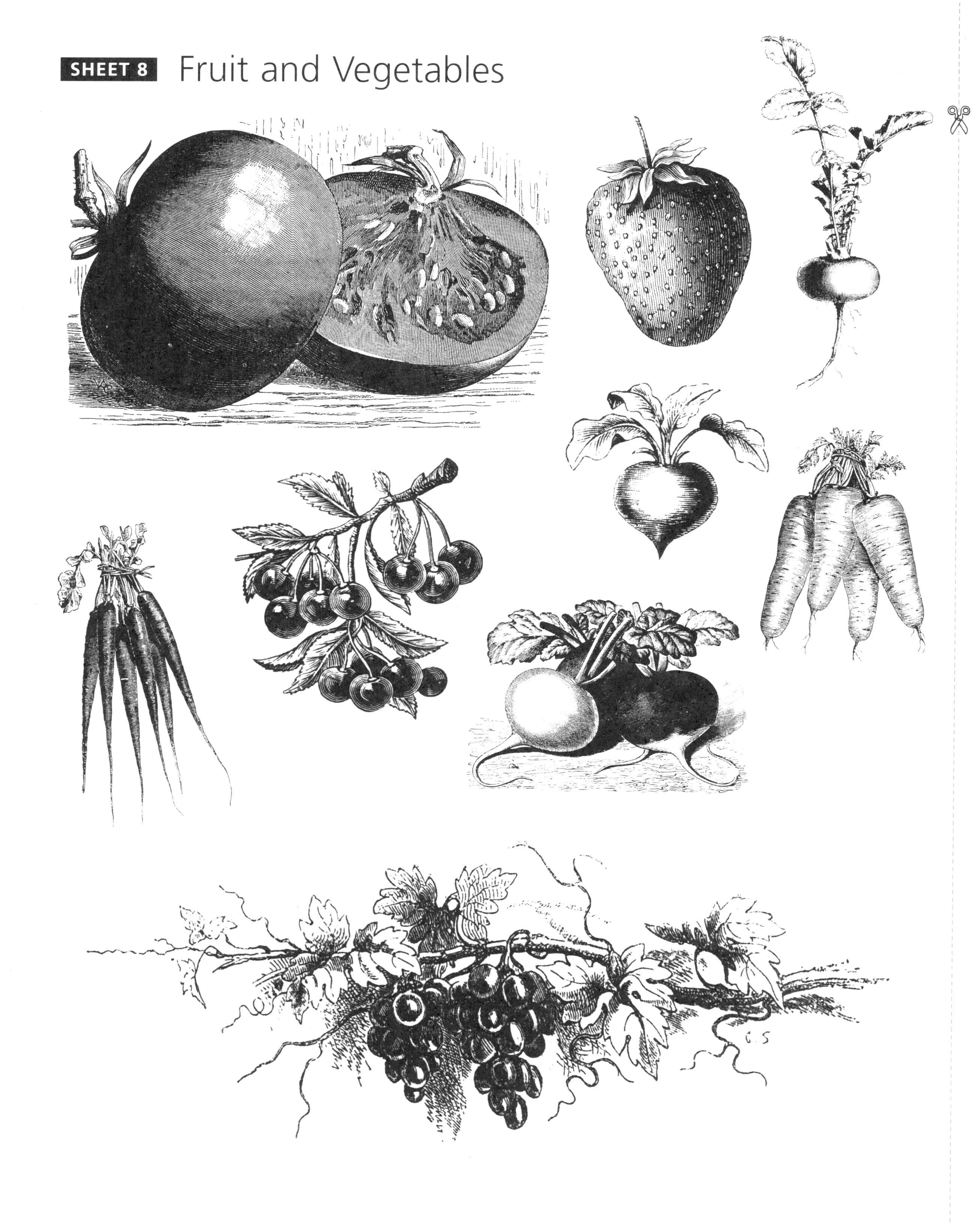

SHEET 9 Food

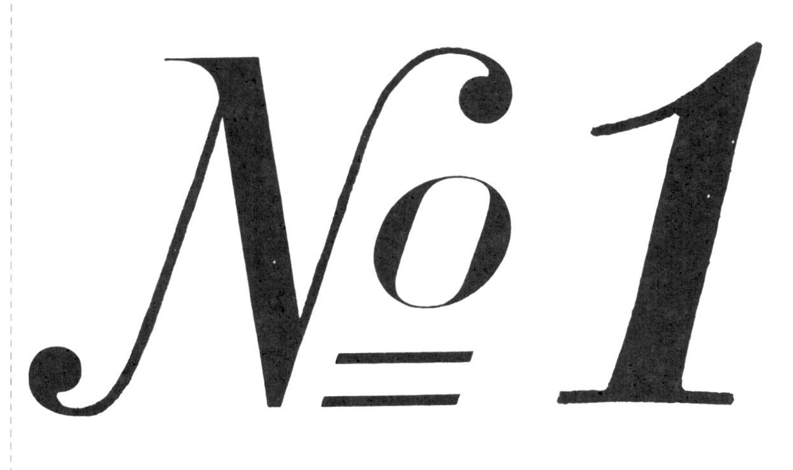

SHEET 10 Flowers

SHEET 11 Flowers and Foliage

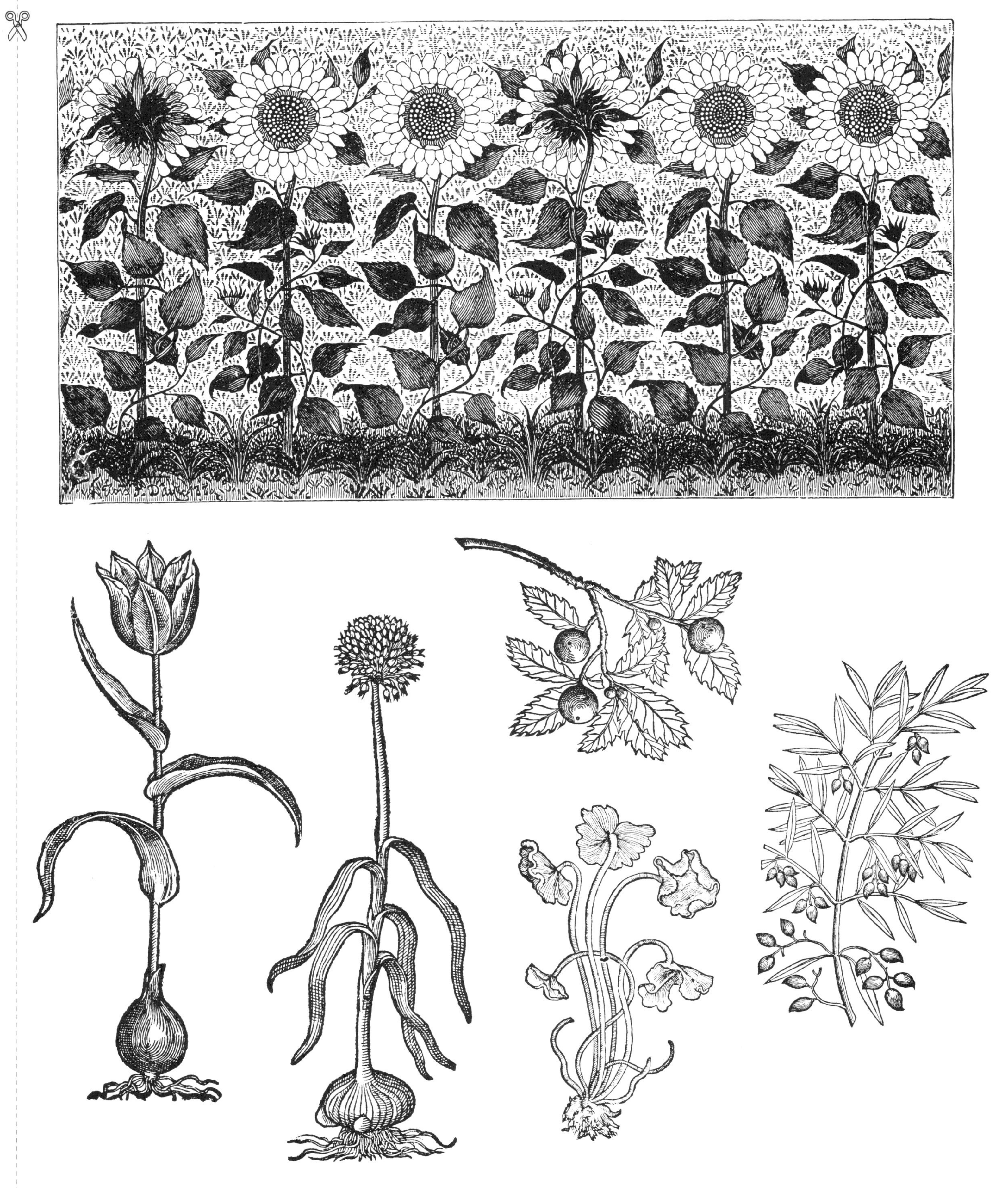

Fish and Shells

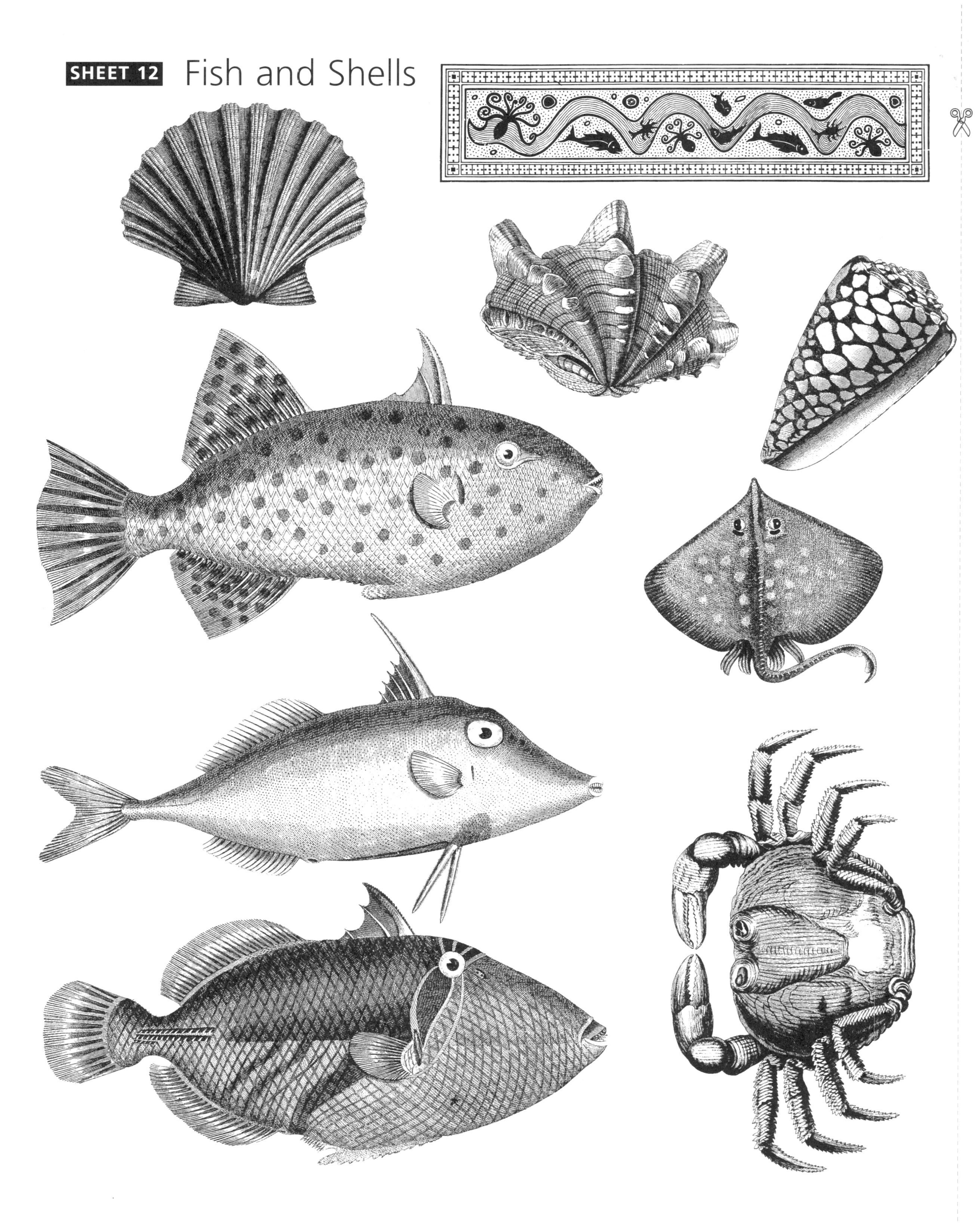

SHEET 13 Birds

SHEET 14 Insects

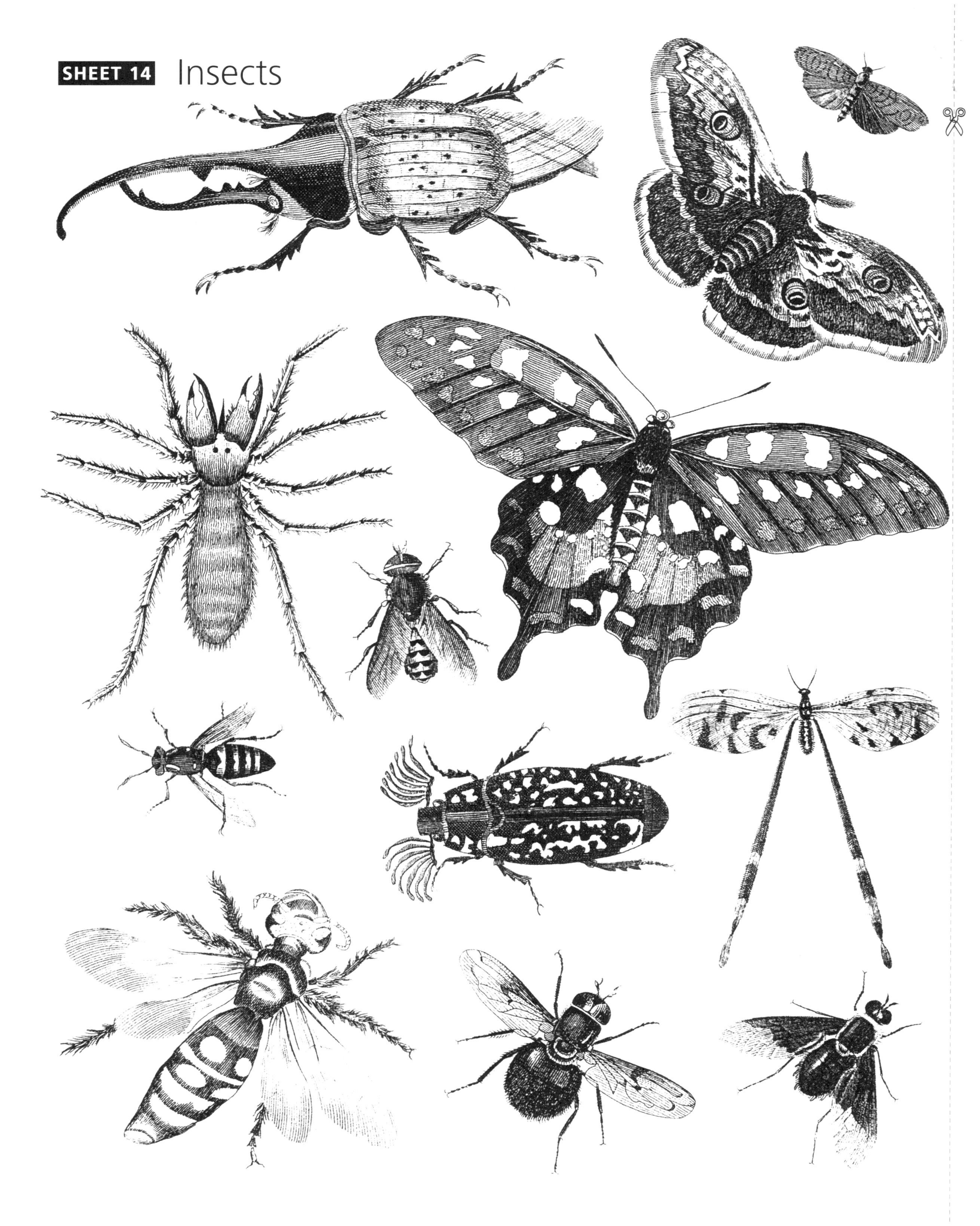

SHEET 15 Alphabets

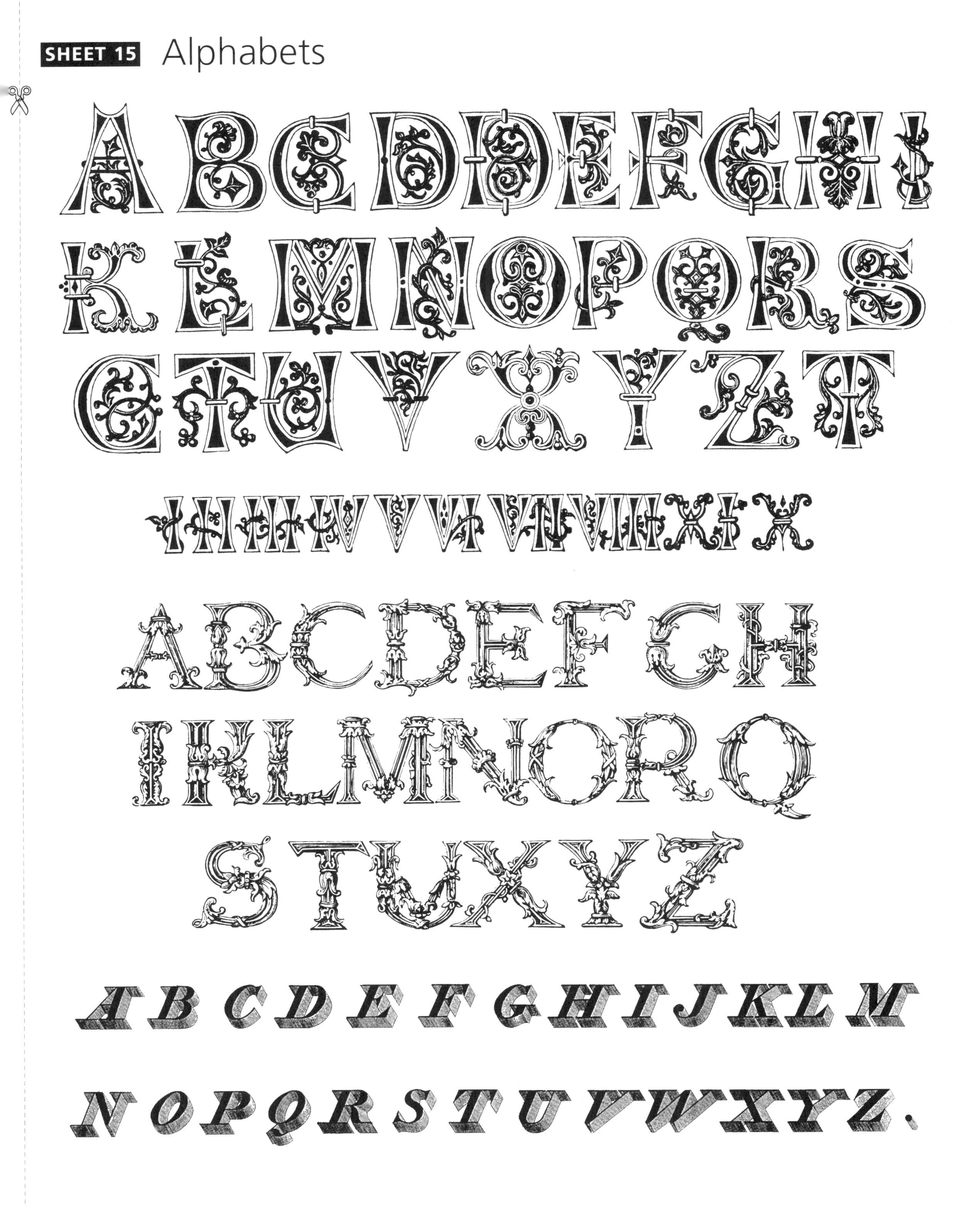

SHEET 16 Clocks and Watches

SHEET 17 Circus

SHEET 18 Circus

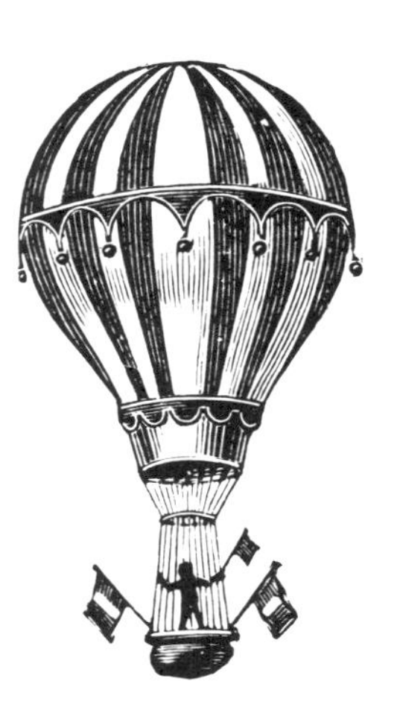

SHEET 19 Children's Games

Sports and Games

SHEET 21 First Aid

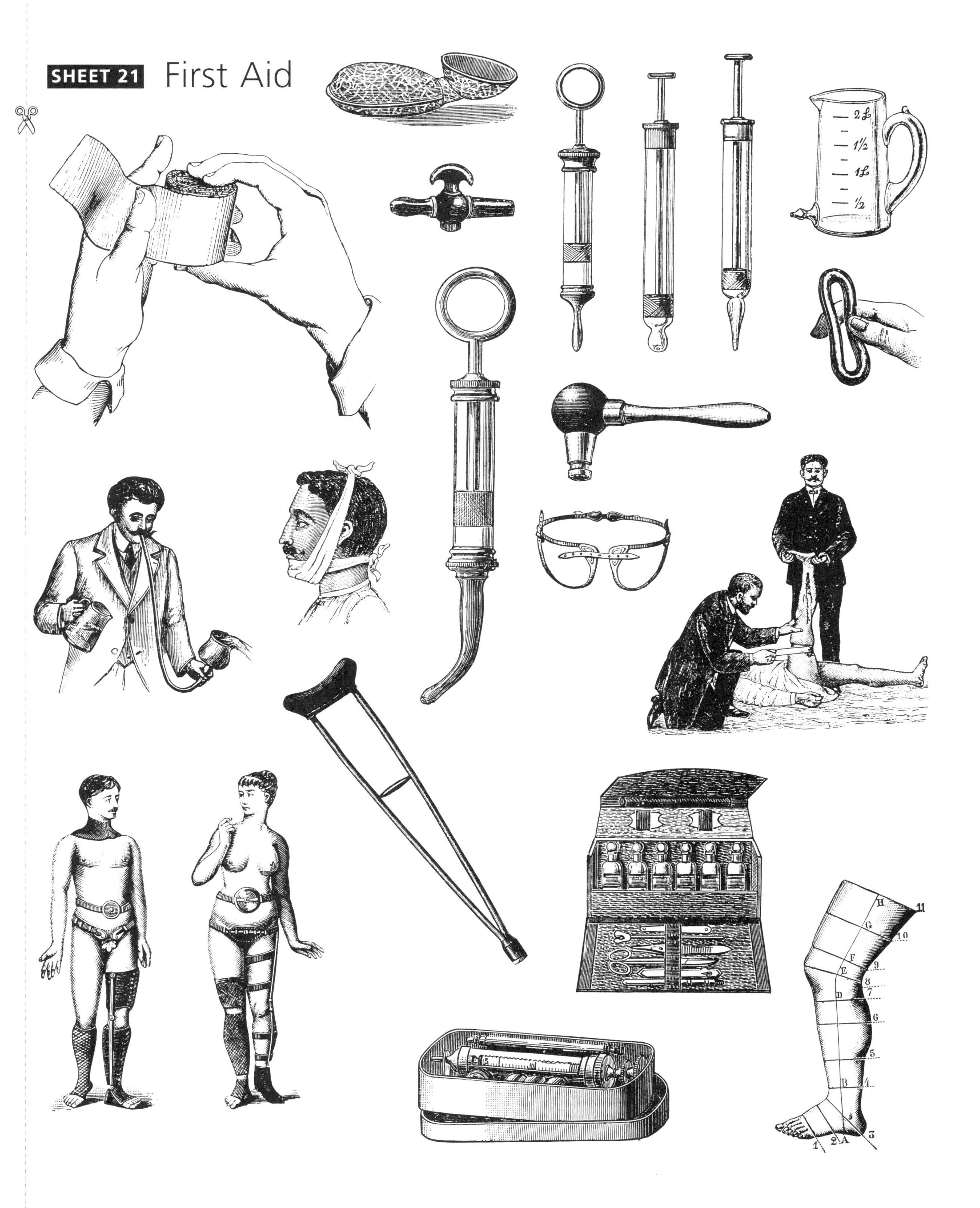

SHEET 22 Kitchen Implements

SHEET 23 China and Cutlery

SHEET 24 Gardening Tools

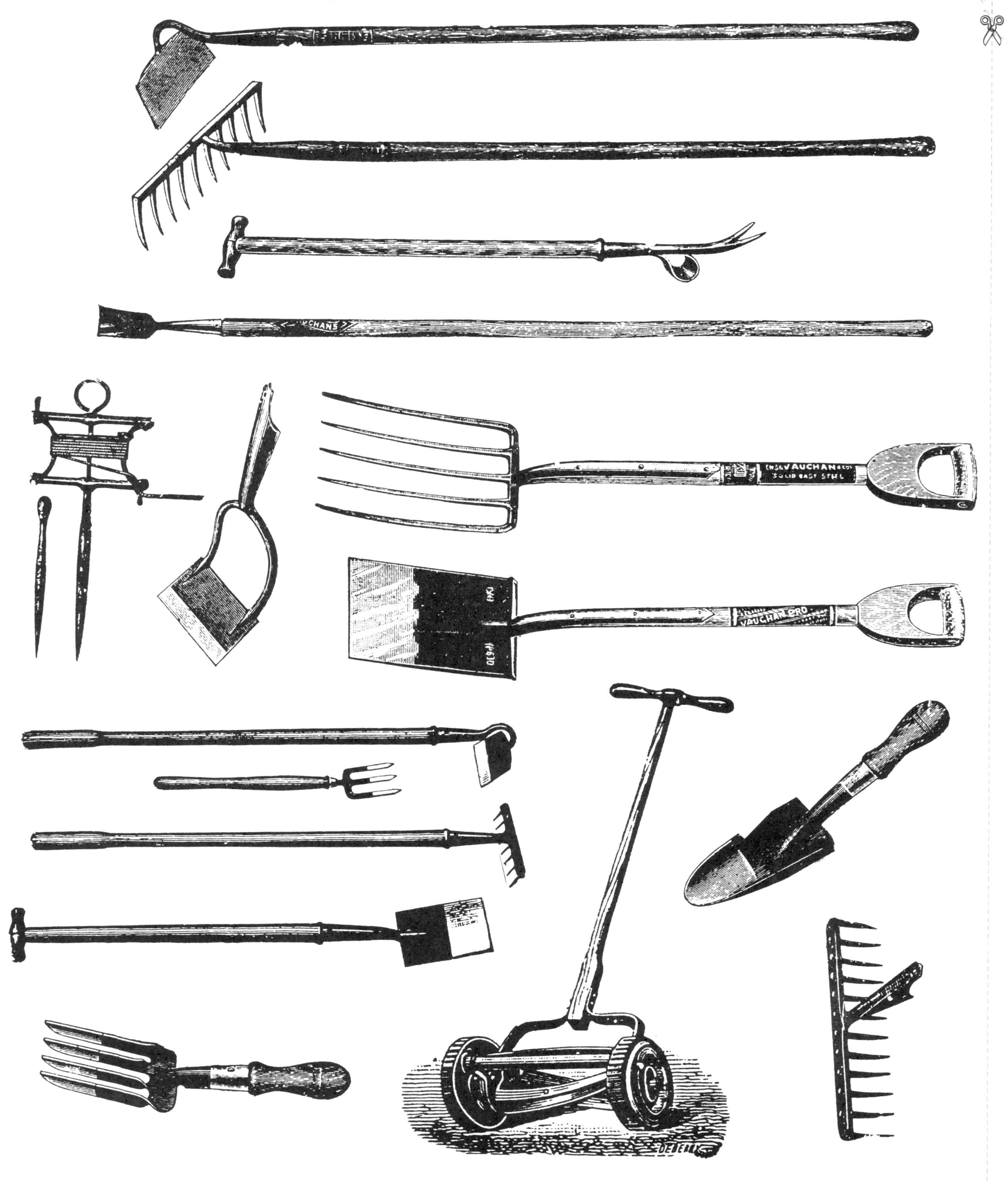

Furniture

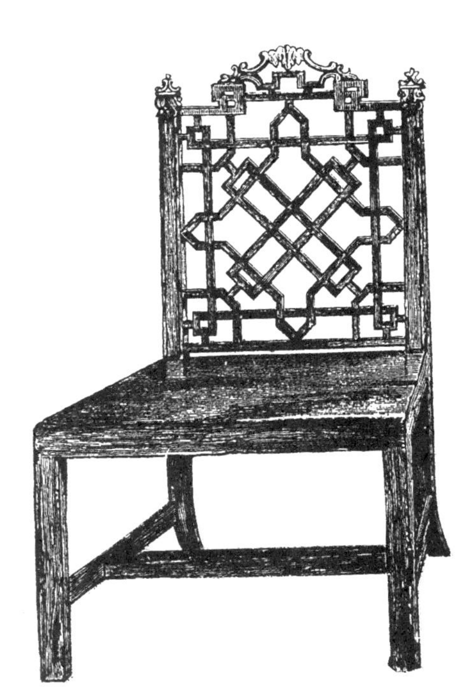

SHEET 26 Musical Instruments

SHEET 27 Household Goods

SHEET 28 Clothes and Accessories

SHEET 29 Architectural Details

SHEET 30 Architectural Details

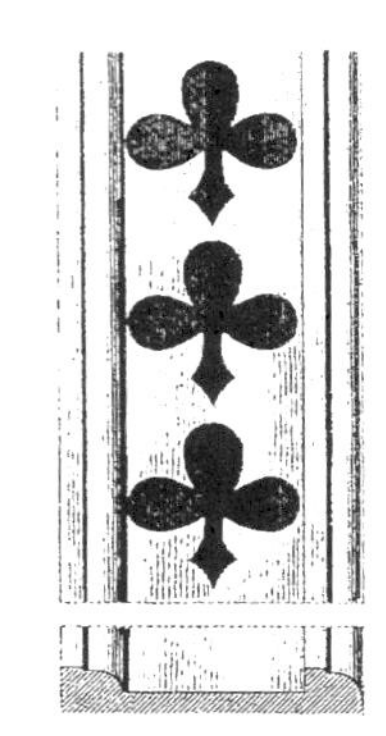

SHEET 31 Architectural Details

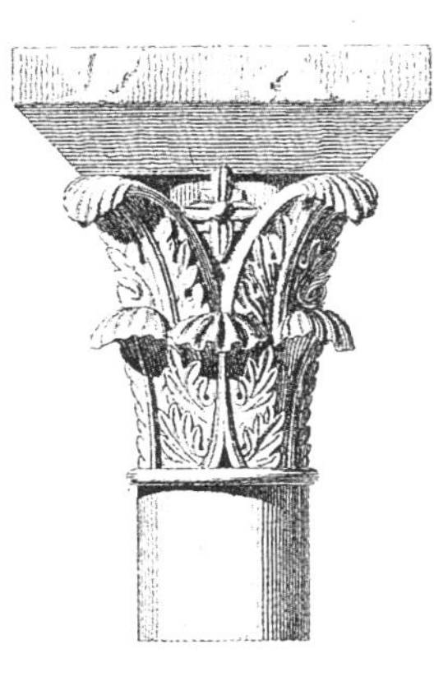

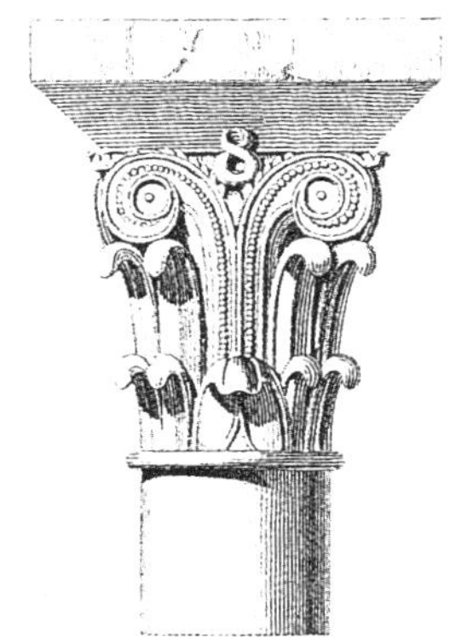

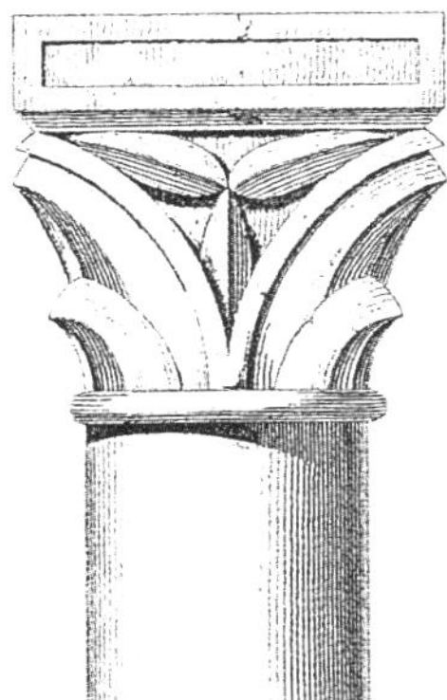

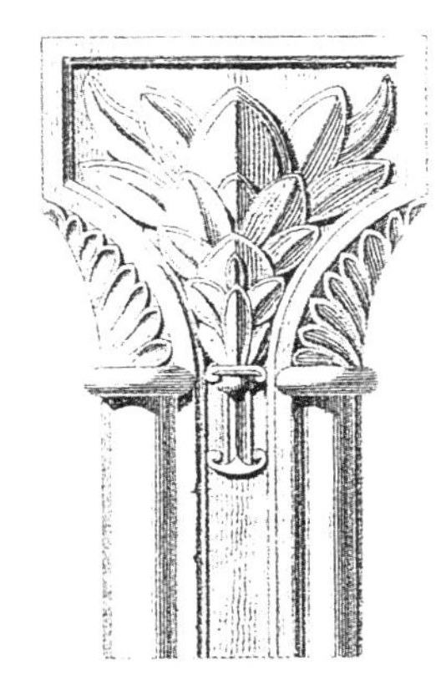

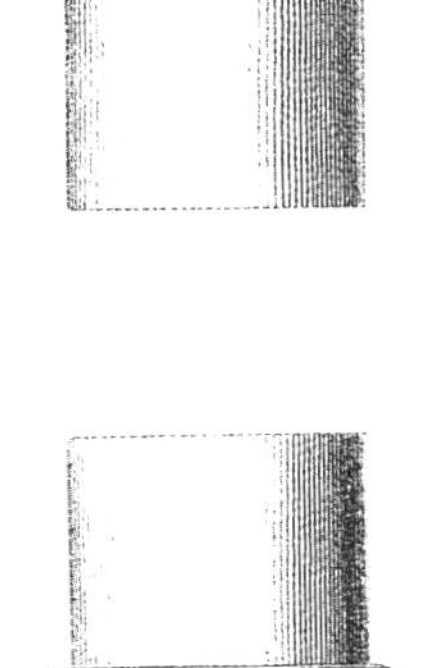

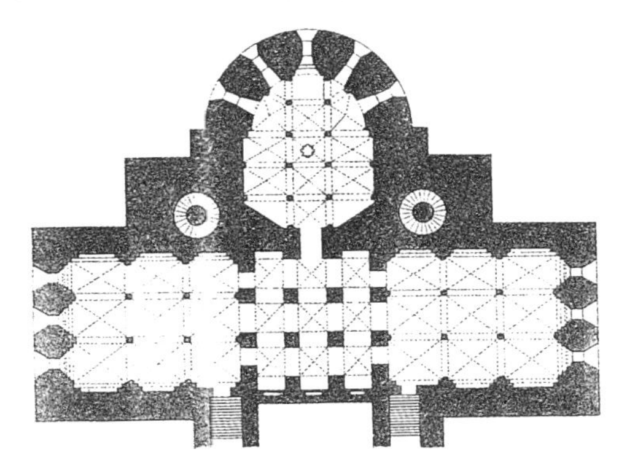

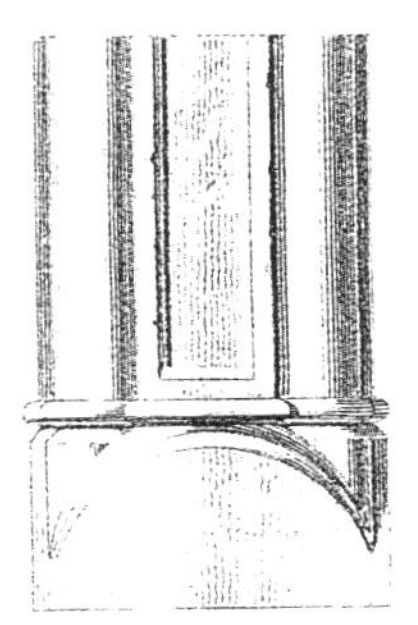